T0202770

Communications
in Computer and Information Science   2069

Rationale

The CCIS series is devoted to the publication of proceedings of computer science conferences. Its aim is to efficiently disseminate original research results in informatics in printed and electronic form. While the focus is on publication of peer-reviewed full papers presenting mature work, inclusion of reviewed short papers reporting on work in progress is welcome, too. Besides globally relevant meetings with internationally representative program committees guaranteeing a strict peer-reviewing and paper selection process, conferences run by societies or of high regional or national relevance are also considered for publication.

Topics

The topical scope of CCIS spans the entire spectrum of informatics ranging from foundational topics in the theory of computing to information and communications science and technology and a broad variety of interdisciplinary application fields.

Information for Volume Editors and Authors

Publication in CCIS is free of charge. No royalties are paid, however, we offer registered conference participants temporary free access to the online version of the conference proceedings on SpringerLink (http://link.springer.com) by means of an http referrer from the conference website and/or a number of complimentary printed copies, as specified in the official acceptance email of the event.

CCIS proceedings can be published in time for distribution at conferences or as post-proceedings, and delivered in the form of printed books and/or electronically as USBs and/or e-content licenses for accessing proceedings at SpringerLink. Furthermore, CCIS proceedings are included in the CCIS electronic book series hosted in the SpringerLink digital library at http://link.springer.com/bookseries/7899. Conferences publishing in CCIS are allowed to use Online Conference Service (OCS) for managing the whole proceedings lifecycle (from submission and reviewing to preparing for publication) free of charge.

Publication process

The language of publication is exclusively English. Authors publishing in CCIS have to sign the Springer CCIS copyright transfer form, however, they are free to use their material published in CCIS for substantially changed, more elaborate subsequent publications elsewhere. For the preparation of the camera-ready papers/files, authors have to strictly adhere to the Springer CCIS Authors' Instructions and are strongly encouraged to use the CCIS LaTeX style files or templates.

Abstracting/Indexing

CCIS is abstracted/indexed in DBLP, Google Scholar, EI-Compendex, Mathematical Reviews, SCImago, Scopus. CCIS volumes are also submitted for the inclusion in ISI Proceedings.

How to start

To start the evaluation of your proposal for inclusion in the CCIS series, please send an e-mail to ccis@springer.com.

Taye Girma Debelee · Achim Ibenthal ·
Friedhelm Schwenker ·
Yehualashet Megersa Ayano
Editors

Pan-African Conference on Artificial Intelligence

Second Conference, PanAfriCon AI 2023
Addis Ababa, Ethiopia, October 5–6, 2023
Revised Selected Papers, Part II

 Springer

Editors
Taye Girma Debelee (iD)
Ethiopian Artificial Intelligence Institute
Addis Ababa, Ethiopia

Friedhelm Schwenker (iD)
Universität Ulm
Ulm, Germany

Achim Ibenthal (iD)
HAWK University of Applied Sciences
and Arts
Göttingen, Germany

Yehualashet Megersa Ayano (iD)
Ethiopian Artificial Intelligence Institute
Addis Ababa, Ethiopia

ISSN 1865-0929   ISSN 1865-0937 (electronic)
Communications in Computer and Information Science
ISBN 978-3-031-57638-6   ISBN 978-3-031-57639-3 (eBook)
https://doi.org/10.1007/978-3-031-57639-3

This Springer imprint is published by the registered company Springer Nature Switzerland AG
The registered company address is: Gewerbestrasse 11, 6330 Cham, Switzerland

Paper in this product is recyclable.

Preface

This edition presents the proceedings of the *Pan-African Conference on Artificial Intelligence 2023 (PanAfriCon AI 2023)*. Starting in 2022, this annual conference focuses on African AI developments. The high demand for such a platform can be seen from a tripling of contributions just within one year. At the same time AI is developing new worldwide trends at an accelerating pace. Examples are generative AI and many professional applications in medical AI, agriculture, autonomous maneuvering, financial technologies, cyber security, office applications, and many more. Related to the African continent, more and more countries are developing AI strategies. It is the set goal of this conference to be an arena for the exchange of best practices and the establishment of joint Pan-African efforts to provide solutions for Africa's key twenty-first century challenges in the social, economic, and ecologic domains.

PanAfriCon AI 2023 aimed at bringing together AI researchers, computational scientists, engineers, entrepreneurs, and decision-makers from academia, industry, and government institutions to discuss the latest trends, opportunities, and challenges of the application of AI in different sectors of the continent. During the conference, attendees were able to exchange the latest information on techniques and workflows used in artificial intelligence in a variety of research fields.

After issuing the call for proposals for the conference in May 2023, 134 contributions were received by August 15, 2023. Based on a single-blind review by 2 reviewers, 71 of the 134 contributions were accepted for presentation and for the further publication process. Following an update considering reviewer comments and Springer CCIS series requirements, these contributions were single-blind reviewed by 3 peer reviewers per paper, out of which 26 submissions were finally accepted for publication in 2 volumes, Springer CCIS 2068 and 2069. The first volume covers medical AI, natural language processing, and text and speech processing, the second AI in finance and cyber security, autonomous vehicles, AI ethics, and life sciences.

Due to logistical considerations, the conference was held completely virtually, organized by the headquarters of the Ethiopian Artificial Intelligence Institute EAII in Addis Ababa. The conference was opened by the director of the EAIC, Worku Gachena. In a keynote speech, Achim Ibenthal, HAWK University of Applied Sciences and Arts, Germany, talked on the elements of a Pan-African AI strategy, spanning a wide range from African history to value chain aspects, opportunities and threats of AI technology. Taye Girma, deputy general director of the research and development cluster of the EAII concluded the conference opening with a session briefing. Four parallel sessions were conducted on October 5 and 6, covering AI in health, AI in services, AI in cyber security, and natural language processing.

This conference would not have been possible without the help of many people and organizations. First of all, we are grateful to all the authors who submitted their contributions. We thank the members of the program committee and the peer reviewers for performing the task of selecting contributions for conference presentations and proceedings with all due diligence.

Finally we hope that readers may enjoy the selection of papers and get inspired by these excellent contributions.

February 2024

Taye Girma Debelee
Achim Ibenthal
Friedhelm Schwenker
Yehualashet Megersa Ayano

Address of the Director General EAII

Worku Gachena Negera
Director General
Ethiopian Artificial Intelligence Institute (EAII)
Addis Ababa
Ethiopia

It is with immense pride and a profound sense of purpose that we present this compilation of papers from the second PanAfriCon AI conference, hosted by our institution, the Ethiopian Artificial Intelligence Institute. Our second conference, conducted virtually in October 2023, brought together a diverse group of thinkers, innovators, and practitioners from across the globe, united in their quest to expand the horizons of artificial intelligence (AI). The thematic areas for this year's proceedings are both a reflection of our current priorities and a forecast of the trajectory AI is poised to take. Natural language processing (NLP) stands at the forefront, exemplifying our commitment to breaking down barriers in communication and making technology accessible to diverse linguistic cultures. The papers in this domain showcase advancements that are not only technical marvels but also bridges to inclusivity. AI in medicine highlights the strides we are making in personalized healthcare, predictive diagnostics, and treatment plans tailored by intelligent systems that learn and adapt. These papers underscore the life-saving potential of AI, offering insights into a future where medicine is intimately informed by machine learning algorithms. The intersection of AI and finance is dissected, revealing how artificial intelligence is reshaping everything from everyday banking to complex investment strategies. The collection presents a visionary perspective on how AI is becoming indispensable in navigating the complexities of the financial world. AI's role in cybersecurity is increasingly critical, and the contributions here detail innovative approaches to thwarting digital threats. These papers are a testimony to AI's evolving capability to act as a guardian in the digital realm, where security is paramount. Autonomous vehicles, an exhilarating field, promise to redefine mobility. The research presented provides a glimpse into the sophisticated AI systems that pilot this transformative technology, ensuring safety and efficiency. Lastly, ethical AI, perhaps the most crucial of all themes, binds the others. The discourse on ethics in AI is not just about guiding principles for AI development but also about the very fabric of the society we aspire to build with these tools. These proceedings are more than just a collection of academic papers; they are a beacon of hope and a map to a future where AI serves humanity, amplifies our potential, and addresses the challenges that span our complex world. As we stand on the cusp of an AI-augmented age, the Ethiopian AI Institute is proud to have been the convener of such a monumental exchange of knowledge. It is my hope that the insights within these pages will inspire and challenge the reader to engage with AI in ways that are profound, positive, and transformative.

Organization

Ethiopian Artificial Intelligence Institute, Addis Ababa, Ethiopia

General Chairs

Taye Girma Debelee	Ethiopian Artificial Intelligence Institute, Ethiopia
Achim Ibenthal	HAWK Univ. of Applied Sciences & Arts, Germany
Friedhelm Schwenker	Ulm University, Germany
Yehualashet Megersa Ayano	Ethiopian Artificial Intelligence Institute, Ethiopia

Program Committee

Fitsum Assamnew	Addis Ababa Institute of Technology, Ethiopia
Abeba Birhane	Mozilla Foundation, San Francisco, Trinity College, Dublin, Ireland
Bisrat Derebessa	Addis Ababa Institute of Technology, Ethiopia
Biniam Gebru	North Carolina A&T State University, USA
Taye Girma Debelee	Ethiopian Artificial Intelligence Institute, Ethiopia
Beakal Gizachew	Addis Ababa Institute of Technology, Ethiopia
Achim Ibenthal	HAWK Univ. of Applied Sciences & Arts, Germany
Worku Jiffara	Adama Science & Technology University, Ethiopia
Solomon Kassa	1888EC, Ethiopia
Yehualashet Megersa Ayano	Ethiopian Artificial Intelligence Institute, Ethiopia
Thomas Meyer	University of Cape Town, South Africa
Samuel Rahimeto	Ethiopian Artificial Intelligence Institute, Ethiopia
Friedhelm Schwenker	Ulm University, Germany
Bruce Watson	Stellenbosch University, South Africa
Teklu Urgessa	Adama Science & Technology University, Ethiopia

Peer Reviewers

The editors and organizing committee sincerely thank the following peer reviewers:

Mesfin Abebe	Adama Science & Technology University, Ethiopia
Allan Anzagira	JPMorgan Chase & Co., USA
Natnael Argaw	Addis Ababa Institute of Technology, Ethiopia
Fitsum Assamnew	Addis Ababa Institute of Technology, Ethiopia
Clayton Baker	University of Cape Town, South Africa
Yohannes Bekele	North Carolina A&T State University, USA
Bisrat Bekele Ergecho	Addis Ababa Institute of Technology, Ethiopia
Sinshaw Bekele Habte	HAWK Univ. of Applied Sciences & Arts, Germany
Victoria Chama	University of Cape Town, South Africa
Bisrat Derebessa	Addis Ababa Institute of Technology, Ethiopia
Biniam Gebru	North Carolina A&T State University, USA
Fraol Gelana	Ethiopian Artificial Intelligence Institute, Ethiopia
Frances Gilles-Webber	University of Cape Town, South Africa
Taye Girma Debelee	Ethiopian Artificial Intelligence Institute, Ethiopia
Beakal Gizachew	Addis Ababa Institute of Technology, Ethiopia
Robbel Habtamu	Addis Ababa Institute of Technology, Ethiopia
Sintayehu Hirphasa	Adama Science & Technology University, Ethiopia
Achim Ibenthal	HAWK Univ. of Applied Sciences & Arts, Germany
Worku Jiffara	Adama Science & Technology University, Ethiopia
Gopi Krishina	Adama Science & Technology University, Ethiopia
Amanuel Kumsa	Ethiopian Artificial Intelligence Institute, Ethiopia
Louise Leenen	University of the Western Cape, South Africa
Surafel Lemma	Addis Ababa Institute of Technology, Ethiopia
Adane Leta	University of Gonder, Ethiopia
Yehualashet Megersa Ayano	Ethiopian Artificial Intelligence Institute, Ethiopia
Tilahun Melak	Adama Science & Technology University, Ethiopia
Chala Merga	Addis Ababa Institute of Technology, Ethiopia
Fitsum Mesfine	Ethiopian Artificial Intelligence Institute, Ethiopia
Thomas Meyer	University of Cape Town, South Africa
Daniel Moges	Ethiopian Artificial Intelligence Institute, Ethiopia
Sudhir Kumar Mohapatra	Sri Sri University, India
Henock Mulugeta	Addis Ababa Institute of Technology, Ethiopia
Abdul-Rauf Nuhu	North Carolina A&T State University, USA

Srinivas Nune	Adama Science & Technology University, Ethiopia
Samuel Rahimeto	Ethiopian Artificial Intelligence Institute, Ethiopia
Friedhelm Schwenker	Ulm University, Germany
Bahiru Shifaw	Adama Science & Technology University, Ethiopia
Ram Sewak Singh	Adama Science & Technology University, Ethiopia
Solomon Teferra	Addis Ababa University, Ethiopia
Menore Tekeba	Addis Ababa Institute of Technology, Ethiopia
Natnael Tilahun	Addis Ababa Science and Technology University, Ethiopia
Tullu Tilahun	Addis Ababa Science and Technology University, Ethiopia
Rosa Tsegaye	Ethiopian Artificial Intelligence Institute, Ethiopia
Amin Tuni	Arsi University, Ethiopia
Teklu Urgessa	Adama Science & Technology University, Ethiopia
Steve Wang	University of Cape Town, South Africa
Bruce Watson	Stellenbosch University, South Africa
Degaga Wolde	Ethiopian Artificial Intelligence Institute, Ethiopia
Leul Wuletaw	University of Michigan, USA
Martha Yifiru	Addis Ababa University, Ethiopia
Azmeraw Yotorawi	Addis Ababa Institute of Technology, Ethiopia

Submission Platform

Friedhelm Schwenker	Ulm University, Germany
Yehualashet Megersa Ayano	Ethiopian Artificial Intelligence Institute, Ethiopia

Collaborating Partner

HAWK University of Applied Sciences and Arts, Göttingen, Germany

Contents – Part II

AI in Finance and Cyber Security

Improving the Accuracy of Financial Bankruptcy Prediction Using
Ensemble Learning Techniques 3
   Anthonia Oluchukwu Njoku, Berthine Nyunga Mpinda,
   and Olushina Olawale Awe

Deep Learning and Machine Learning Techniques for Credit Scoring:
A Review 30
   Hana Demma Wube, Sintayehu Zekarias Esubalew,
   Firesew Fayiso Weldesellasie, and Taye Girma Debelee

Refining Detection Mechanism of Mobile Money Fraud Using MoMTSim
Platform 62
   Denish Azamuke, Marriette Katarahweire, Joshua Muleesi Businge,
   Samuel Kizza, Chrisostom Opio, and Engineer Bainomugisha

An Investigation and Analysis of Vulnerabilities Surrounding
Cryptocurrencies and Blockchain Technology 83
   Isabelle Heyl, Dewald Blaauw, and Bruce Watson

Towards a Supervised Machine Learning Algorithm for Cyberattacks
Detection and Prevention in a Smart Grid Cybersecurity System 107
   Takudzwa Vincent Banda, Dewald Blaauw, and Bruce W. Watson

Classification of DGA-Based Malware Using Deep Hybrid Learning 129
   Bereket Hailu Biru and Solomon Zemene Melese

A Review and Analysis of Cybersecurity Threats and Vulnerabilities,
by Development of a Fuzzy Rule-Based Expert System 151
   Matida Churu, Dewald Blaauw, and Bruce Watson

Autonomous Vehicles

Neural Network Based Model Reference Adaptive Control of Quadrotor
UAV for Precision Agriculture 171
   Muluken Menebo, Lebsework Negash, and Dereje Shiferaw

AI Ethics and Life Sciences

Systems Thinking Application to Ethical and Privacy Considerations
in AI-Enabled Syndromic Surveillance Systems: Requirements
for Under-Resourced Countries in Southern Africa 197
    Taurai T. Chikotie, Bruce W. Watson, and Liam R. Watson

Fake vs. Real Face Discrimination Using Convolutional Neural Networks 219
    Khaled Eissa and Friedhelm Schwenker

Hybrid of Ensemble Machine Learning and Nature-Inspired Algorithms
for Divorce Prediction 242
    Kalkidan A. Sahle and Abdulkerim M. Yibre

Author Index 265

Contents – Part I

Medical AI

Machine Learning Based Stroke Segmentation and Classification
from CT-Scan: A Survey 3
   Elbetel Taye Zewde, Mersibon Melese Motuma,
   Yehualashet Megersa Ayano, Taye Girma Debelee,
   and Degaga Wolde Feyisa

Multitask Deep Convolutional Neural Network with Attention
for Pulmonary Tuberculosis Detection and Weak Localization
of Pathological Manifestations in Chest X-Ray 46
   Degaga Wolde Feyisa, Yehualashet Megersa Ayano,
   Taye Girma Debelee, and Samuel Sisay Hailu

Automated Kidney Segmentation and Disease Classification Using
CNN-Based Models 60
   Akalu Abraham, Misganu Tuse, and Million Meshesha

Generating Synthetic Brain Tumor Data Using StyleGAN3 for Lower
Class Enhancement 73
   Ahmed Abdalaziz and Friedhelm Schwenker

Optimized Machine Learning Models for Hepatitis C Prediction:
Leveraging Optuna for Hyperparameter Tuning and Streamlit for Model
Deployment 88
   Uriel Nguefack Yefou, Pauline Ornela Megne Choudja, Binta Sow,
   and Abduljaleel Adejumo

Explainable Rhythm-Based Heart Disease Detection from ECG Signals 101
   Dereje Degeffa Demissie and Fitsum Assamnew Andargie

Development of an Explainable Heart Failure Patients Survival Status
Prediction Model Using Machine Learning Algorithms 117
   Betimihirt Getnet Tsehay Demis and Abdulkerim M. Yibre

Natural Language Processing, Text and Speech Processing

Transfer of Models and Resources for Under-Resourced Languages
Semantic Role Labeling 141
   Yesuf Mohamed and Wolfgang Menzel

Speaker Identification Under Noisy Conditions Using Hybrid Deep
Learning Model 154
    Wondimu Lambamo, Ramasamy Srinivasagan, and Worku Jifara

State-of-the-Art Approaches to Word Sense Disambiguation:
A Multilingual Investigation 176
    Robbel Habtamu and Beakal Gizachew

Ge'ez Syntax Error Detection Using Deep Learning Approaches 203
    Habtamu Shiferaw Asmare and Abdulkerim Mohammed Yibre

Tigrinya End-to-End Speech Recognition: A Hybrid Connectionist
Temporal Classification-Attention Approach 221
    Bereket Desbele Ghebregiorgis, Yonatan Yosef Tekle,
    Mebrahtu Fisshaye Kidane, Mussie Kaleab Keleta,
    Rutta Fissehatsion Ghebraeb, and Daniel Tesfai Gebretatios

DNN-Based Supervised Spontaneous Court Hearing Transcription
for Amharic 237
    Martha Yifiru Tachbelie, Solomon Teferra Abate, Rosa Tsegaye Aga,
    Rahel Mekonnen, Hiwot Mulugeta, Abel Mulat, Ashenafi Mulat,
    Solomon Merkebu, Taye Girma Debelee, and Worku Gachena

Typewritten OCR Model for Ethiopic Characters 250
    Bereket Siraw Deneke, Rosa Tsegaye Aga, Mesay Samuel, Abel Mulat,
    Ashenafi Mulat, Abel Abebe, Rahel Mekonnen, Hiwot Mulugeta,
    Taye Girma Debelee, and Worku Gachena

Compressed Amharic Text: A Prediction by Partial Match
Context-Modeling Algorithm 262
    Yalemsew Abate Tefera, Tsegamlak Terefe Debella,
    Habib Mohammed Hussien, and Dereje Hailemariam Woldegebreal

Author Index 281

AI in Finance and Cyber Security

Improving the Accuracy of Financial Bankruptcy Prediction Using Ensemble Learning Techniques

Anthonia Oluchukwu Njoku[1](✉) , Berthine Nyunga Mpinda[1] ,
and Olushina Olawale Awe[1,2]

[1] African Institute for Mathematical Sciences, Limbe, Cameroon
`anthonia.njoku@aims-cameroon.org`, `bmpinda@aimsammi.org`
[2] State University of Campinas, Sao Paulo, Brazil
`oawe@unicamp.br`

Abstract. Financial institutions have been seeking ways to improve their bankruptcy prediction capabilities to mitigate the disruptive effects of future bankruptcies. One such way is using machine learning models. However, financial datasets are often imbalanced, posing a significant challenge for building effective predictive models. In this work, three resampling techniques are used to produce the datasets that were used for model building: oversampling, undersampling, and hybrid sampling. We evaluate the effectiveness of these sampling techniques on five machine learning models (Logistic Regression, Bagging, Random Forest, Support Vector Machine, Neural Networks) in predicting financial bankruptcies. We also investigate the impact of ensembling on model performance by stacking the high-performing individual models using a logistic regression meta-classifier. Our results show that hybrid sampling provides a better balance of accuracy and accountability for the minority (bankrupt) class, which makes it a suitable balancing technique for imbalanced financial datasets. Additionally, ensembling the models using stacking improved the performance of the models, resulting in a better performance for predicting bankruptcies. Remarkably, our proposed model demonstrated an outstanding accuracy of 99.75% while models from existing literature, and previous studies reported accuracies ranging from 83% to 98% for similar ensemble stacking tasks. Results from this study will be useful for practitioners in the finance sphere in making informed decisions, managing risks and choosing the right models for bankruptcy prediction.

Keywords: Bankruptcy Prediction · Imbalanced Data · Ensemble Stacking

1 Introduction

In recent years, bankruptcies have become common in the business world, with many companies experiencing financial difficulties and ultimately failing [12]. This trend is attributed to the growing complexity of modern economies and the intense

T. G. Debelee et al. (Eds.): PanAfriConAI 2023, CCIS 2069, pp. 3–29, 2024.
https://doi.org/10.1007/978-3-031-57639-3_1

competition that businesses face. The Enron scandal serves as a vivid example of how competition can impact a company's financial health. Once the 7th largest company in the US, Enron faced heightened competition in its operations, leading to shrinking profits and mounting losses that ultimately led to its bankruptcy [9]. In the current business environment, it is crucial to make profitable economic decisions that benefit all stakeholders. One crucial factor in making such decisions is the ability to predict bankruptcy in a company. Accurately predicting bankruptcy can help to manage risks, prevent financial losses, and enhance decision-making. Identifying financial distress before it gets too severe is essential for shareholders, creditors, and regulators to steer a company to a better future with informed decisions [23]. Prediction of Bankruptcies is thus a crucial area of research in the financial and accounting sectors as it seeks to identify early warning signals to prevent financial losses and economic instability [30].

This subject has also not only been a critical issue for both financial institutions and investors, it has also significant implications for achieving Sustainable Development Goals (SDGs). The ability to predict the possibility of a company's bankruptcy can help in its sustainability as it enables stakeholders to mitigate risks and prevent financial losses. This ultimately leads to the advancement of sustainable economic growth which SDG 9 - Industry, Innovation, and Infrastructure aims to achieve [26]. Bankruptcy prediction has been studied extensively for several decades, and various methods have been proposed to address this problem. These methods can be broadly grouped into two types: traditional statistical methods and machine learning methods. Traditional statistical methods rely on linear regression, discriminant analysis, and other statistical techniques to identify significant financial ratios that can predict bankruptcy [3,4,22]. On the other hand, machine learning methods use advanced algorithms such as artificial neural network (ANN), decision trees (DT), and support vector machines (SVM) to learn from historical data and predict future outcomes [17,29,37].

While traditional statistical methods have been used successfully in predicting bankruptcy, they have several limitations [16]. This is largely due to their dependence on statistical techniques. For instance, they rely on assumptions of linear separability, independence of explanatory variables, and normality of variables, which are often not the case when financial ratios are used as predictor variables [16]. Also, there is the limitation of high dimension, as most traditional statistical models can only analyze a small number of predictors [20]. Machine learning methods, on the other hand, can address these limitations by automatically learning from data without relying on predefined assumptions or financial ratios. Machine learning algorithms can perform on large volumes of data to identify non-linear relationships that may not be captured by traditional statistical methods. Moreover, machine learning algorithms can adapt to changing environments, making them more robust and flexible than traditional statistical methods. Machine learning models perform with approximately 10% more accuracy in relation to traditional models [7]. By thus utilizing this technology and innovation, financial institutions and investors can have a more efficient and effective approach to managing risks, promoting economic growth, and creating job opportunities (SDG 8 - Decent Work and Economic Growth).

We focus on the prediction of financial bankruptcy in this paper using a stacked ensemble of machine-learning algorithms. While typical machine learning employs the use of one model to learn from training data and make predictions, this is not the case with ensemble learning. Ensemble learning is a powerful technique that combines the predictions of several models to achieve better prediction performance and robustness [33]. Stacking is an ensemble learning method that enables the combination of several models, utilizing a meta-classifier. We also introduce the concept of hybrid sampling which is a balancing technique that combines oversampling and undersampling methods for managing data imbalance. This research analyzes the financial ratios of multiple companies and utilizes a stacked ensemble of diverse machine-learning models to predict their likelihood of experiencing bankruptcy. The goal is to enhance economic decision-making by providing businesses with more informed choices, thereby improving their ability to make profitable decisions. This research's findings will provide noteworthy enlightenment for financial institutions and investors in making informed decisions and managing risks, ultimately contributing to the advancement of the financial industry and the achievement of SDGs 8 and 9.

2   Review of Related Works

The prediction of corporate bankruptcy holds immense economic significance, as it affects numerous stakeholders. This has prompted various countries to accord considerable attention to the issue and to devise legislation to govern companies and regulate bankruptcies. For instance, in Poland, bankruptcy is governed by the Bankruptcy Law Act of February 28, 2003 [1]. This law specifies that bankruptcy can only be declared in relation to a debtor who is insolvent, meaning that they are unable to meet their financial obligations as they come due or that their liabilities exceed the value of their assets [19]. There has also been an increasing body of research carried out in this regard, in the bid to find techniques and methodologies for accurate and usable predictions. Fitzpatrick conducted one of the earliest studies on the prediction of bankruptcy in 1932 [24]. Fitzpatrick [15], in his work, identified worthy-to-note differences between financially sound companies and otherwise distressed companies. He put forward the idea that there exist five distinct phases that ultimately culminate in business failure [5]. Subsequently, Altman in 1968, devised the Z-score model which employs five financial ratios as a means of anticipating firm bankruptcies in the United States. He utilized Multiple Discriminant Analysis (MDA) techniques to assess the likelihood of bankruptcy for a sample of companies. The approach he employed relied on the assumption that variables adhere to a normal distribution. However, subsequent researchers challenged this assumption and instead advocated for the multiformity of variables [37].

Balcaen & Ooghe [6] in their paper gave an extensive discourse regarding the utilization of traditional statistical methods for predicting business failure. They provided detailed insights into four specific approaches and how they can be applied in corporate failure prediction. The MDA technique was also used with Logistic Regression to develop distinct models for predicting bankruptcy specifically for manufacturing companies in the Slovak Republic [32]. The findings obtained from the models indicated that the logit model demonstrated better prediction accuracy. To overcome the limitation of MDA in handling only a finite number of predictors, Jones [20], employed the gradient boosting model in his research. The aforementioned model successfully handled a substantial number of predictors and effectively ranked them, according to their predictive capability. Kitowski et al. [24], in their article, aimed to validate the reliability of specific discriminant and logit models. They suggested utilizing a wide range of these models to evaluate bankruptcy risk, along with other available approaches for estimating the financial state of an enterprise, such as risk scoring methods and traditional ratio analysis.

In response to the limitations associated with traditional statistical methods, an increasing number of researchers endeavored to find solutions by applying machine learning techniques instead. Barboza et al. [7] in their study employed four machine learning models to forecast bankruptcy in companies one year before its actual occurrence. Comparing the performance of these models with results from logistic regression, neural networks, and discriminant analysis and it was discovered that the models used outperform the other techniques. In comparing the best models, using a testing sample with all explanatory variables, the random forest model yielded an 87% accuracy, while the logistic regression and linear discriminant analysis models yielded 69% and 50% accuracy respectively. The contribution of this model is such that in the application, it would alert the companies one year ahead of time to retrace and make better informed decisions that could actually avert the impending bankruptcy. Qu et al. [34] in their paper, reviewed some machine learning and deep learning techniques that over the years have been used, and Son et al. [38] in their study focused on managing the characteristic skewness in financial data. They employed a model that showed a 17% average improvement over existing models. They also addressed the problem of interpretability by analyzing the importance of features identified by the Extreme Gradient Boosting (XGBoost) model. SVM and ANN were evaluated as suitable classification methods for the prediction of bankruptcy [17]. It was noted that even though the SVM performed better, it was not as applicable in practice as the neural structured model was.

Whilst researchers built models based on machine learning techniques, Alam et al. [2] carried out a study that significantly improved predictive accuracy. They proposed the use of machine learning techniques with other economic ratios, as well as Altman's Z-score variables in relation to liquidity, profitability, solvency, and leverage to build an efficient model. Results from this research gave very impressive accuracies with random forest outperforming other techniques with a 99% accuracy. As research in the field progressed, Kou et al. [25] introduced a new technique for Small and Medium Enterprises (SMEs), a new technique which involved using data inferred from transaction history and other payment network-based variables instead of conventional accounting-based financial ratios.

They proposed a feature-selection approach to optimize the data's dimension and enhance classification performance.

The findings demonstrated that the model incorporating payment variables and variables of the transactional data outperformed other classifiers in terms of economic benefit. The research highlighted the potential of integrating payment and transactional data-based features to enhance model performance, particularly with regard to AUC and profit. Thilakarathna et al. [39] used 29 financial ratios, that have not been used in previous work related to bankruptcy prediction. These predictors provided an avenue for a novel deep neural network to be built. The results from the model reveal the possibility of a better and earlier prediction of bankruptcy compared to the approaches from previous studies. Working with only three financial ratios that could be easily gotten, Shetty et al. [37] used machine learning models to predict bankruptcies with an accuracy range of 82%–83%. Their model proposed a simple and user-friendly tool to predict bankruptcies.

Several researchers have concentrated on merging the benefits of various machine learning models in a technique called ensemble learning in order to increase prediction performance. This method has been used by a few researchers in building high-performance models for different fields. Kim & Kang [23] proposed a neural network ensemble to improve the performance of existing neural networks when used for bankruptcy prediction tasks. Results after experiments on Korean enterprises revealed that neural networks that were bagged and boosted outperformed traditional neural networks. The ensemble technique was also used for the detection of hate speech in the Indonesian language [14]. The proposed model used five base classification algorithms, as well as two ensemble approaches, hard and soft voting. The findings of the experiment demonstrate that using the ensemble method can increase classification performance. Soft voting produced the best results, with F1 measures of 79.8% on unbalanced datasets and 84.7% on balanced datasets. Ensembling techniques were also used for reproducible medical image classification in the work of Muller et al. [33]. A pipeline was proposed for analyzing the performance of some ensemble techniques: Stacking, Augmenting, and Bagging. The pipeline includes cutting-edge preprocessing and picture augmentation techniques, as well as nine robust convolution neural network topologies. It was tested against four widely used datasets on medical imagery of various complexity. In addition, twelve pooling algorithms for aggregating many different predictions were investigated, ranging from less complex statistical functions such as unweighted averaging to more complicated ones such as SVMs. From the results, stacking achieved the highest performance boost, with up to a 13% rise in F1-score. Augmenting demonstrated constant improvement of about 4% and is also relevant to pipelines based on a single model. Using cross-validation bagging revealed a considerable performance boost comparable to stacking, resulting in an F1-score rise of up to 11%.

In their study, Hosseini et al. [18] aimed to predict Peak Particle Velocity (PPV), an undesirable consequence that occurs during vibration emission in surface mines' blasted benches. They employed a stacked ensemble model to aggregate the results of some top-performing models. The findings indicated that the ensemble model significantly improved the accuracy of PPV prediction compared to the best individual models, achieving an impressive accuracy of 98.2%. In the study of Learning et al. [27], they applied stacking techniques to improve the accuracy of Twitter sentiment classification. They constructed an ensemble by combining three algorithms: SVM, K-Nearest Neighbors (KNN), and C5.0. The ensemble was further enhanced by employing two meta-learners, Random Forest (RF) and Logistic Regression. Notably, the stacking-based ensemble comprising SVMRadial, KNN, and C5.0 with RF as the meta-classifier achieved an accuracy of 86.7%.

Buyrukoglu and Savas [11] tackled the task of determining and classifying the positions of football players in their study. They employed a stacking-based ensemble machine learning model utilizing the FIFA'19 game dataset. To optimize performance, they selected 10 features using four distinct feature selection algorithms. As for the individual base algorithms, Deep Neural Networks, RF, and Gradient Boosting were employed, with Logistic Regression serving as the meta-learner in the stacking-based model. The findings indicated that the utilization of the Chi-square feature selection technique, together with the stacking-based ensemble learning model gave the highest accuracy, reaching 83.9%.

This study adds to the existing body of literature in the field as it proposes an ensemble stacking technique based on five state-of-the-art models: SVM, neural networks, logistic regression, random forest, bagging for improved performance in the prediction of bankruptcy on the Polish bankruptcy dataset.

3   Study Design and Research Methods

Here, we present the methodology used for conducting this study. It outlines the research design, data preparation techniques, as well as machine learning models that were employed to address and achieve the objectives of the study. By providing a detailed description of the methodology, this section aims to enhance the credibility and transparency of the research findings and enable readers to assess the validity and reliability of the study. Figure 1 introduces the approach that we follow through the study: 1) data preparation; 2) feature selection: 3) training some base classifiers; and 4) ensemble stacking.

3.1   Data Acquisition and Preprocessing

In this step, data preparation refers to preprocessing the dataset for modelling. We present a dataset that consists of financial ratios from multiple Polish companies, along with a label indicating if the company was bankrupt or not. The data set contains 43,405 samples and 64 features and was collected over a period of five years. For use with base models, the data had to undergo several processing stages as it was originally untidy. Preprocessing this dataset involved filling the missing values in the dataset, selecting important features and balancing the data to avoid biases in model training.

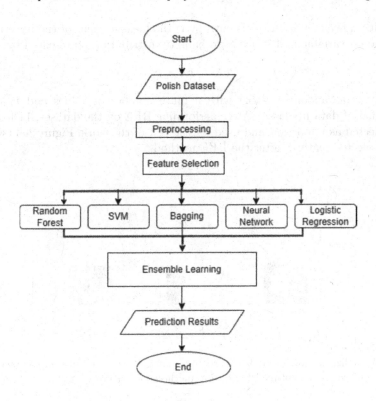

Fig. 1. Bankruptcy Prediction Flowchart.

Multiple Imputation by Chained Equations (MICE) technique was used to handle missing data in this study. This approach is widely utilized for handling missing values in datasets. It performs under the assumption that the missing data occur randomly (MAR) and that the likelihood of a variable being missing depends on the observed data [21]. The basic idea behind MICE is to fill in the values that are missing with plausible estimates, relying on the available information in the data set. MICE accomplishes this by creating separate regression models for each variable with missing values, and then using these models to generate multiple imputed data sets.

Recursive Feature Elimination (RFE) was also employed for selecting features in this study. It is a popular method for feature selection in machine learning. It involves removing attributes recursively and building a model on the remaining attributes until the needed number of features is attained. This algorithm begins by creating a model of the whole variables and allocating an importance score to each predictor [10]. The model is then rebuilt with the least important predictor(s) eliminated, and the significance scores are computed again until all the features have been explored. The RFE's stability largely depends on the base model which is used for the iteration. One of such underlying models is a simple Decision Tree (DT) algorithm [28]. It computes the dataset's information entropy and splits it layer by layer. This entropy is the measurement of the uncertainty of a random variable and it has a probability distribution represented by:

$$P(X = x_i) = p_i, \tag{1}$$

where X is a random variable with a finite number of values and p_i is the proportion of data in class i. After performing RFE on the dataset, 11 features were selected as important and used for the rest of the work. Figure 2 shows the plot of selected features after the RFE method.

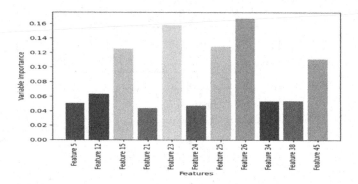

Fig. 2. Visualization of selected features with their level of feature importance after performing Recursive Feature Elimination method for feature selection.

One of the primary preparation steps in this work was to manage the imbalance in the dataset, which, if not managed properly, could lead to bias in model training because machine learning models tend to favour classes with the most representation. The imbalance ratio of the dataset before balancing was 0.0506, which means that for every five samples in the minority class, there are 100 more in the majority. The highly imbalanced nature of the data therefore made it imperative for balancing to be carried out. Figure 3 represents an illustration of the class distribution in the original dataset.

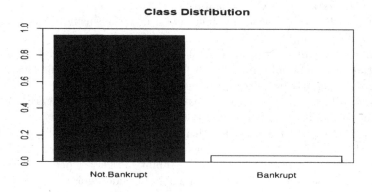

Fig. 3. Class Distribution of the Polish dataset.

To balance the minority class in this study, we use the oversampling, under-sampling and hybrid sampling methods: In oversampling, synthetic samples are generated for the minority class by sampling from a normal distribution with estimates of the mean and covariance gotten from the minority class. The original dataset and the synthetic samples are then combined to create the oversampled dataset. Table 1 represents the data's class distribution after oversampling was performed.

Table 1. Class Distribution of Oversampled Data.

Class	Count
Bankrupt	28876
Not Bankrupt	28974

The under-sampling method is used in balancing the majority class by select-ing a random subset of instances from the majority class without replacement. Equal number of instances are selected as in the minority class. Mathematically, let's say we have a dataset D with n instances and m features. We also have a binary target variable y with two classes, where the positive class is the minority class. We denote the positive class by C^+ and the negative class by C^-.

Let n^+ be the number of instances in the minority class C^+ in D, and let n^- be the number of instances in the majority class C^- in D. We randomly select n^+ instances from C^- without replacement to create the balanced dataset D. The n^+ instances from C^+ and the n^+ instances selected from C^- are then combined to create the balanced dataset D. The class distribution of the dataset after undersampling is shown in Table 2

Table 2. Class Distribution of Undersampled Data.

Class	Count
Bankrupt	1506
Not Bankrupt	1519

Combining random oversampling and random undersampling can yield inter-esting results as the classifier's performance could be improved to a greater extent [31]. In hybrid sampling, the oversampling method is used to balance the minor-ity class, and then the under-sampling method is used to balance all the data. After hybrid sampling was carried out, Table 3 shows the class representation of the dataset.

Table 3. Class distribution for Hybrid sampling.

Class	Count
Bankrupt	15183
Not Bankrupt	15297

3.2   Ensemble Stacking

Bankruptcy prediction as a machine learning problem entails classifying companies based on their financial status - whether they are bankrupt or not. This task falls under the category of classification problems, which employ supervised learning algorithms for accurate classification. In supervised machine learning, the model is trained on labelled data automatically and improves with experience without being explicitly programmed. The datasets used in supervised learning contain a label feature that is dependent on the other features in the datasets. Through this, the machine learns the relationship between the input features and the label feature, allowing it to make adequate predictions for future data.

In this paper, an ensemble stacking algorithm was used. Ensemble stacking is a machine learning technique used to combine multiple models in order to improve prediction accuracy. It combines the base machine learning models to perform better than the best classifier [27]. Suppose X is a training set and N be the number of base models you want to use in your ensemble. The models are trained on the dataset to generate N different sets of predictions. These predictions are combined using a meta-model, which takes the N predictions as inputs and generates a final prediction. In the case where the meta-model uses linear regression, it is formulated thus:

$$y = w_0 + \sum_{i=1}^{n} w_i p_i. \tag{2}$$

where: •y is the final prediction

•$w_0, w_1, ..., w_n$ are weights to be learned during training, and

•$p_1, p_2, ..., p_n$ are the N sets of predictions from the individual models.

Once the meta-model is trained, it can be used to generate predictions on new data by first using the individual models to generate predictions, and then passing those predictions through the meta-model.

We consider the following base models used in this work: Logistic regression is a statistical model that is used in predictive analysis and machine learning. It is especially widely-used when the label is binary due to its simplicity and effectiveness [8]. It is used for estimating the probability of an event's occurrence based on one or more predictor variables. Its core, the sigmoid (logit) function, is employed to transform linear regression into logistic regression, making it more appropriate for addressing binary classification problems [36].

Given a training dataset consisting of n observations, each with m predictor variables and a binary outcome variable, we can represent the predictor variables for the ith observation as a column vector x_i:

$$x_i = [x_{1i}, x_{2i}, \dots, x_{mi}]^T, \tag{3}$$

where the first element of x_i is 1, which corresponds to the intercept term β_0 in the logistic regression model. This model can be represented as:

$$p(y_i = 1 | x_i, \beta) = \frac{1}{(1 + \exp(-x_i^T \beta))}, \tag{4}$$

where $\beta = [\beta_0, \beta_1, \beta_2, \dots, \beta_m]^T$ is the vector of coefficients that we want to estimate.

Random Forest is a robust algorithm for machine learning tasks. By combining numerous decision trees, it creates a powerful model. The algorithm randomly selects data subsets and features to build individual decision trees, and the final prediction is aggregated from all the trees. Random Forest performs well on large datasets due to its foundation in decision tree theory. However, it is susceptible to overfitting [40]. The Random Forest algorithm has several hyperparameters that can be tuned to improve its performance, such as the number of trees, the number of features to select at each split, and the maximum depth of the decision trees. These hyperparameters can be selected by techniques such as cross-validation or grid search.

Neural networks are a type of machine learning model that are used to approximate complex functions, classify data, and make predictions. They consist of interconnected layers of processing units, or neurons, that can learn to recognize patterns and relationships in the data they are trained on. Unlike traditional linear models that have certain limitations, neural networks offer distinct advantages in capturing nonlinear behavior in economics and financial markets [35]. This makes them particularly valuable in these domains, as they can effectively model complex and non-linear relationships that exist within these systems.

Let x be the input data matrix of size $m \times n$, where m is the number of training examples and n is the number of input features. The input data is fed forward through the network, and the network produces an output y, which can be a class label or a set of values. Neural networks are comprised of different layers of interconnected nodes, which apply a series of linear and nonlinear transformations to the input data in order to produce an output. The neurons in a layer are connected to the neurons in the previous and next layers by weights, which are learnable parameters that are adjusted during training. The vectors of the input features and the weights are respectively $x = [x_1, x_2, \dots, x_n]$ and $w = [w_1, w_2, \dots, w_n]$. The dot product of the input and weights give:

$$x.w = (x_1 \times w_1) + (x_2 \times w_2) + \dots + (x_n \times w_n) = \sum_{i=1}^{n} x_i w_i. \tag{5}$$

To generate the required output, we include a bias term, b, which allows the activation function to move to the left or the right. Then, we have:

$$z = wx + b, \tag{6}$$

where z is the weighted sum of the inputs plus a bias b.

The output of each neuron is then passed through a nonlinear activation function $f(i,j)$, at the jth neuron in the ith layer, to introduce nonlinearity into the model. The neural network produces its output through the final layer, which typically employs a distinct activation function compared to the other layers. For instance, in binary classification tasks, a sigmoid activation function may be used, while multi-class classification problems often employ softmax activation.

During training, the weights and biases are iteratively adjusted using an optimization algorithm to minimize a loss function $L(y, y_{pred})$ that measures the disparity across the computed output y_{pred} and the actual output y. Backpropagation is utilized to compute the gradient of the loss function with respect to the weights and biases, enabling their update in subsequent iterations:

$$w^{(k+1)} = w^{(k)} - \gamma \frac{\partial L}{\partial w}, \tag{7}$$

$$b^{(k+1)} = b^{(k)} - \gamma \frac{\partial L}{\partial b}, \tag{8}$$

with $10^{-6} < \gamma < 1$ as the learning rate. The learning rate is the determinant of how much both the weights and biases will be updated with each iteration, and the gradients are computed using the chain rule. Once the neural network is trained, it is then used to get predictions on new data by passing the data through the network and computing the output using the learned weights and biases.

Bagging, which stands for bootstrap aggregating, is a popular ensemble method used to improve the accuracy and robustness of models. The working principle of bagging involves training several models on different subsets of the training data and subsequently aggregating their predictions to derive a final prediction. The decision tree classifier is a very common base classifier used in bagging to train subsets of the dataset. Bagging can also be viewed as an approximation of the Bayes optimal classifier [13], where each model m_j is trained on a subset of the data and makes independent predictions. By combining the predictions of all models, we can obtain a more accurate and robust prediction than any single model.

Support Vector Machine (SVM) is a supervised learning model suitable for regression and classification problems. It excels in classifying binary labelled problems, where the goal is to separate data points into two distinct classes. The fundamental concept of SVM involves identifying a decision boundary, referred to as a hyperplane, which maximizes the margin between the two classes [29]. The margin represents the distance between the decision boundary and the nearest data point from either class. In this mathematical description, we will focus on the case of binary classification.

Let us assume that we have a training dataset consisting of input-output pairs, denoted as:

$$D = \{(x_1, y_1), (x_2, y_2), \ldots, (x_n, y_n)\}, \tag{9}$$

where x_i in Eq. 9 is the input vector and y_i is the corresponding output, which takes on one of two possible values, either -1 or $+1$. We proceed to find a hyperplane in the feature space that separates the two classes with the largest possible margin. The hyperplane's equation in an n-dimensional space can be represented in the equation:

$$w^T x + b = 0, \tag{10}$$

where w is a vector normal to the hyperplane and b is the bias term. The distance between a point x_i and the hyperplane can be measured as:

$$d(x_i) = \frac{|w^T x_i + b|}{||w||}, \tag{11}$$

where $||w||$ is the Euclidean norm of w.

The SVM algorithm finds the hyperplane by solving the following optimization problem:

$$\min_w \frac{1}{2} ||w||_2 \tag{12}$$

subject to

$$y_i(w^T x_i + b) \geq 1, \ \forall i = 1, \ldots, n. \tag{13}$$

Various techniques, such as quadratic programming or gradient descent, can be employed to solve this optimization problem. By obtaining the solution to this problem, we can determine the values of w and b that define the hyperplane. In situations where the data is not linearly separable, the SVM algorithm can still be applied by introducing a slack variable ξ_i for each training example x_i. This adjustment transforms the optimization problem into the following form:

$$\min_w \frac{1}{2} ||w||_2 + C \sum_{i=1}^{n} \xi_i \tag{14}$$

subject to

$$y_i(w^T x_i + b) \geq 1 - \xi_i, \tag{15}$$

and

$$\xi_i \geq 0, \forall i = 1, \ldots, n. \tag{16}$$

Here, C is a parameter that controls the trade-off between maximizing the margin and minimizing the misclassification errors. The slack variable ξ_i measures the degree of misclassification of the ith training example. Finally, once the SVM algorithm has learned the hyperplane, it can be used to predict the class of a new input vector x by evaluating the sign of $w^T x_i + b$. If the result is positive, the input belongs to the positive class, otherwise it belongs to the negative class.

3.3 Performance Evaluation Metrics

A Confusion matrix is an essential tool for evaluating how well a model has performed in classification problem. In the context of a binary classification problem, the confusion matrix is defined in Table 4.

Table 4. Confusion Matrix for Binary Classification.

	Predicted Positive	Predicted Negative
Actual Positive	TP	FN
Actual Negative	FP	TN

It enables the measurement of true positives (TP), false positives (FP), true negatives (TN), and false negatives (FN). In our study, the 'not bankrupt' class is considered the positive class, and the 'bankrupt' class is considered the negative class.

Furthermore, the confusion matrix serves as a basis for calculating different performance metrics, including accuracy, balanced accuracy, and the Matthews correlation coefficient. These metrics can provide a more comprehensive evaluation of the model's performance than just looking at the overall accuracy.

– **Accuracy** The accuracy metric is a commonly used performance metric for classification problems. It measures the proportion of correct predictions made by a classification model, and is defined as follows:

$$Accuracy = \frac{TP + TN}{TP + TN + FP + FN}, \tag{17}$$

where TP is the number of true positives, TN is the number of true negatives, FP is the number of false positives, and FN is the number of false negatives. The accuracy metric ranges from 0 to 1, with higher values indicating better model performance. While the accuracy metric is a useful metric for evaluating model performance, it should be used with caution, especially in situations where the classes are imbalanced, as it can be misleading. In such cases, it is recommended to use other metrics such as the balanced accuracy or Matthew's correlation coefficient.

– **Balanced Accuracy** The Balanced Accuracy (BA) is a performance metric that takes into account both the sensitivity (true positive rate) and specificity (true negative rate) of a classification model. It is defined as follows:

$$Balanced\ Accuracy = \frac{1}{2} \left(\frac{TP}{TP + FN} + \frac{TN}{TN + FP} \right), \tag{18}$$

The BA metric ranges from 0 to 1, with higher values indicating better model performance. By using the BA metric, we can ensure that our model is performing well not only on one class, but on both classes, and can avoid a situation where a model performs well on one class but poorly on the other.

- **Matthew's Correlation Coefficient** Matthew's Correlation Coefficient (MCC) is a performance metric that incorporates information from true positives, true negatives, false positives, and false negatives. The formula for MCC is as follows:

$$MCC = \frac{TP \times TN - FP \times FN}{\sqrt{(TP+FP)(TP+FN)(TN+FP)(TN+FN)}}, \tag{19}$$

The MCC metric ranges from -1 to 1, with higher values indicating better model performance. A score of 1 indicates a perfect prediction, 0 indicates a random prediction, and -1 indicates a completely wrong prediction. While Balanced Accuracy is a simpler metric that provides a balanced representation of accuracy, it is used particularly in slightly balanced datasets. MCC is more suitable for highly imbalanced datasets or when considering both positive and negative predictions comprehensively.
- **Specificity** Specificity (True Negative rate) is a measurement of the proportion of negatives that are correctly identified. This means the proportion of negatives that were actually predicted negative.
  Specificity is given as:

$$Specificity = TN/(TN + FP), \tag{20}$$

where TN is the number of true negatives, and FP is the number of false positives. This metric is especially useful in checking the performance of the classification in imbalanced datasets, where the minority class is the negative class. This is important because, in the context of imbalanced data, accuracy can be misleading and may not fully capture the true performance of the model.

4 Empirical Evidence

In this section, we elaborate our research result and findings.

1. Which resampling technique, namely oversampling, undersampling, or hybrid sampling, optimally enhances the performance of machine learning models on imbalanced financial datasets for bankruptcy prediction?
2. Does the utilization of ensemble stacking result in an improvement in predictive accuracy for financial bankruptcy compared to individual models?

4.1 RQ1: Evaluating Resampling Techniques for Imbalanced Financial Datasets

To address RQ1, we make a comprehensive evaluation of the performance of distinct resampling techniques applied individually to the machine learning models. The goal is to discern the impact of oversampling, undersampling, and hybrid sampling on the effectiveness of these models when dealing with imbalanced financial datasets.

Results from Models. The data obtained from feature selection was divided into a 70 : 30 ratio, with 70% used for training and 30% used for testing. Only the training data was balanced using each of the resampling methods, resulting in three datasets for the study. The test data was left unbalanced to reflect the real-world scenario of imbalanced data that would be encountered by the models. In order to make accurate comparisons, models were also trained using the original imbalanced training set. Additionally, we implement repeated cross-validation during the training process.

– Results from Imbalanced Data The following confusion matrices were obtained from models trained on the imbalanced dataset.

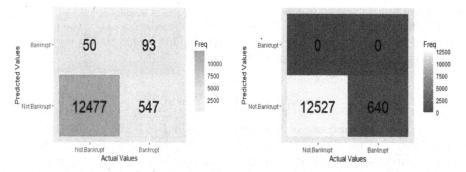

Fig. 4. Confusion Matrix for Imbalanced Bag.

Fig. 5. Confusion Matrix for Imbalanced NN.

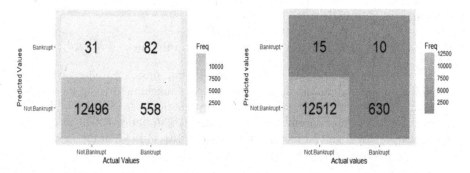

Fig. 6. Confusion Matrix for Imbalanced RF.

Fig. 7. Confusion Matrix for Imbalanced SVM.

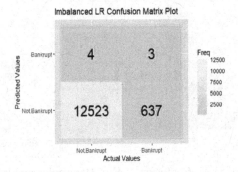

Fig. 8. Confusion Matrix for Imbalanced LR.

From the confusion matrices generated in Figs. 4, 5, 6, 7, and 8, we see the effect of the data imbalance and the necessity of including balancing techniques in the study. The quadrant with the highest count in each matrix, represented by a single differently colored quadrant, indicated the True Positive (TP) quadrant. The SVM model did not classify anything from the negative class, represented by the bankrupt class, while the rest classified very little because they were able to learn to classify the not bankrupt class better. The performance metrics of these models can be seen in Table 5.

Table 5. Performance Metrics of Models with Imbalanced data.

Metrics	SVM	Bagging	NN	RF	LR
Accuracy	0.951	0.955	0.951	0.955	0.951
Sensitivity	0.999	0.996	1.000	0.998	1.000
Specificity	0.016	0.145	0.000	0.128	0.005
MCC	0.071	0.293	0.000	0.293	0.005
F1 Score	0.975	0.977	0.975	0.977	0.975
BA	0.507	0.571	0.500	0.563	0.502

We see that even though the accuracies are good, training with these data will not give the best results from the results we get in specificity, balanced accuracy, and MCC, which are good metrics to use for imbalanced data.
- Results from Undersampling After balancing the data with undersampling techniques, the confusion matrices from the models built can be seen in Figs. 9, 10, 11, 12, and 13.

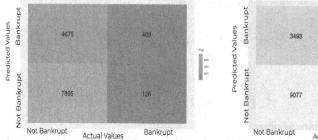

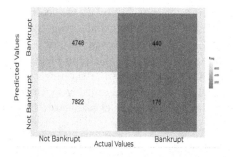

Fig. 9. Undersampled SVM Confusion Matrix.

Fig. 10. Undersampled NN Confusion Matrix.

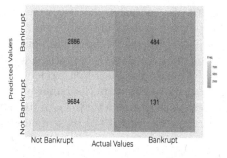

Fig. 11. Undersampled LR Confusion Matrix.

Fig. 12. Undersampled RF Confusion Matrix.

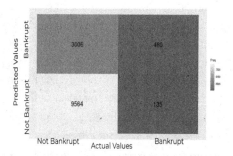

Fig. 13. Undersampled Bag Confusion Matrix.

We see an improvement in the prediction of the bankrupt class than when compared to its predictions with the imbalanced data. This is reflected in the specificity of the models which can be seen in Table 6. The balanced accuracy and MCC also show improvement as undersampling allows the models to take into account both classes. In terms of accuracy, specificity, MCC and BA, random forest performed best with the undersampled data.

Table 6. Performance Metrics of Models with Undersampled data.

Metrics	SVM	Bagging	NN	RF·	LR
Accuracy	0.636	0.762	0.722	0.771	0.627
Sensitivity	0.628	0.761	0.722	0.770	0.622
Specificity	0.795	0.780	0.724	0.787	0.715
MCC	0.183	0.259	0.205	0.269	0.146
F1 Score	0.767	0.859	0.832	0.865	0.761
BA	0.712	0.771	0.723	0.779	0.669

– Results from Oversampling In trying to get the best combination of data and models, we evaluate the models with oversampled data for evaluation of its performance. We see the confusion matrices of these models in Figs. 14, 15, 16, 17, and 18.

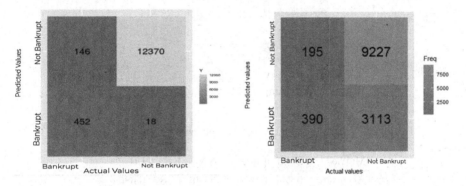

Fig. 14. Oversampled RF Confusion Matrix.

Fig. 15. Oversampled SVM Confusion Matrix.

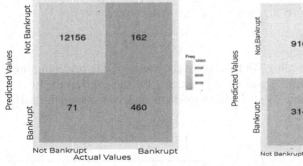

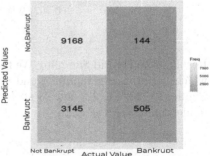

Fig. 16. Oversampled Bag Confusion Matrix.

Fig. 17. Oversampled NN Confusion Matrix.

The matrices in Figs. 14, 15, 16, 17 and 18 show that the data balanced with oversampling method also worked better than the imbalanced data in the classifying and prediction of the bankrupt class. Table 7 shows the performance of each of the models on the data.

Table 7 demonstrates significant enhancements in balanced accuracy, MCC, and specificity for certain models, as compared to the undersampled data. Notably, the bagging and random forest models, both of which are ensemble-based, exhibit notable improvements.

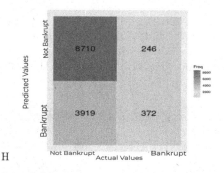

H

Fig. 18. Oversampled LR Confusion Matrix.

Table 7. Performance Metrics of Models with Oversampled data.

Metrics	SVM	Bagging	NN	RF	LR
Accuracy	0.744	0.982	0.746	0.987	0.686
Sensitivity	0.748	0.994	0.745	0.999	0.690
Specificity	0.667	0.739	0.778	0.756	0.602
MCC	0.193	0.791	0.253	0.847	0.131
F1 Score	0.848	0.990	0.848	0.993	0.807
BA	0.707	0.867	0.761	0.877	0.646

– Results from Hybrid Sampling We finally use the method of hybrid sampling, which performs oversampling and then downsamples the oversampled data. Figures 19, 20, 21, 22, and 23 show the confusion matrices gotten from the model built from this data.

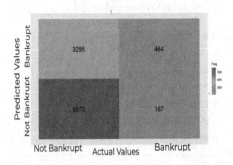

Fig. 19. Hybrid-sampled NN Confusion Matrix.

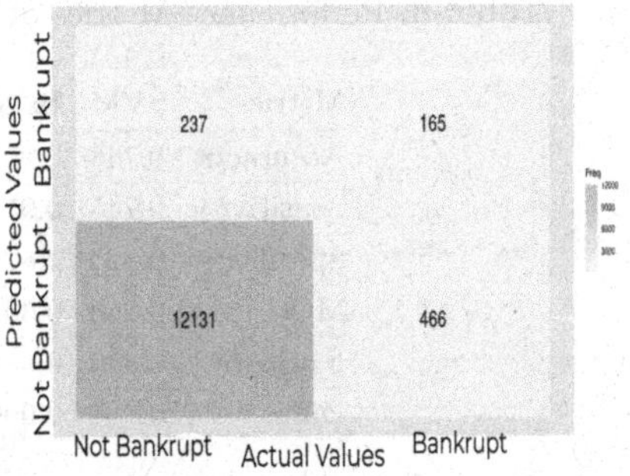

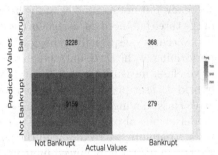

Fig. 20. Hybrid-sampled RF Confusion Matrix.

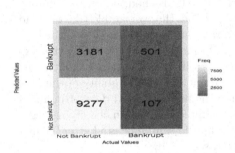

Fig. 21. Hybrid-sampled SVM Confusion Matrix.

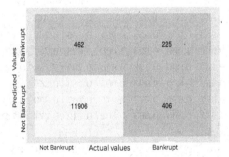

Fig. 22. Hybrid-sampled LR Confusion Matrix.

Fig. 23. Hybrid-sampled Bag Confusion Matrix.

The confusion matrices show that hybrid sampling also does well with prediction of the negative class. Table 8 shows the performance of the models trained on hybrid-sampled data.

Table 8. Performance Metrics of Models with Hybrid-sampled data

Metrics	SVM	Bagging	NN	RF	LR
Accuracy	0.748	0.965	0.748	0.974	0.746
Sensitivity	0.745	0.971	0.747	0.982	0.754
Specificity	0.824	0.829	0.776	0.814	0.577
MCC	0.266	0.307	0.222	0.301	0.150
F1 Score	0.849	0.981	0.850	0.986	0.850
BA	0.784	0.900	0.761	0.898	0.666

Table 9 is the comparison of the 5 models used in this work for all the balancing types. From the results presented in Table 9, it is clear that the models trained on the undersampled dataset had good performance in specificity and balanced accuracy, but not as good as in the hybrid data. It also had the lowest accuracy of the three balanced datasets. On the other hand, the oversampled dataset had good accuracy, especially in bagging and random forest, where it was the best performing of the three datasets, but it performed poorly in detecting the negative class, measured by specificity, except in its neural network model. The hybrid dataset provided a balance of accuracy and accountability of the negative class, having higher accuracies than the undersampled dataset in all its models, while also outperforming the oversampled dataset in classifying negatives. Therefore, we recommend the use of hybrid sampling as a better balancing technique for imbalanced financial datasets.

Table 9. Comparison of 5 models using different data balancing techniques.

Metrics	Imbalanced					Undersampled				
	SVM	BAG	NN	RF	LR	SVM	BAG	NN	RF	LR
ACC	0.951	0.955	0.951	0.955	0.951	0.636	0.762	0.722	0.771	0.627
SENS	0.999	0.996	1.000	0.998	1.000	0.628	0.761	0.722	0.770	0.622
SPEC	0.016	0.145	0.000	0.128	0.005	0.795	0.780	0.724	0.787	0.715
MCC	0.071	0.293	0.000	0.293	0.005	0.183	0.259	0.205	0.269	0.146
F1 Score	0.975	0.977	0.975	0.977	0.975	0.767	0.859	0.832	0.865	0.761
BA	0.507	0.571	0.500	0.563	0.502	0.712	0.771	0.723	0.779	0.669
Metrics	Oversampled					Hybrid				
	SVM	BAG	NN	RF	LR	SVM	BAG	NN	RF	LR
ACC	0.744	0.982	0.746	0.987	0.686	0.748	0.965	0.748	0.974	0.746
SENS	0.748	0.994	0.745	0.999	0.690	0.745	0.971	0.747	0.982	0.754
SPEC	0.667	0.739	0.778	0.756	0.602	0.824	0.829	0.776	0.814	0.577
MCC	0.193	0.791	0.253	0.847	0.131	0.266	0.307	0.222	0.301	0.150
F1 Score	0.848	0.990	0.848	0.993	0.807	0.849	0.981	0.850	0.986	0.850
BA	0.707	0.867	0.761	0.877	0.646	0.784	0.900	0.761	0.898	0.666

4.2 RQ2: Enhancing Predictive Accuracy Through Ensemble Stacking

For RQ2, we investigate whether the collaborative power of ensemble stacking yields a substantial improvement in predictive accuracy compared to the performance of each individual model. Given that the data balanced with Hybrid sampling method performed better than the data balanced with oversampling and undersampling methods, we use it to train the ensemble, stacking four of the high-performing individual models: SVM, bagging, neural networks, and random forest, as presented in Table 10. All models achieved high performance, with accuracy rates of about 90% or higher, except for the neural network which performed at 75%. Finally, the stacked ensemble model gave an accuracy of 99.75%, which was an even better performance than the individual models. In Table 11, we can see that the stacked ensemble was able to account for the bankrupt class in its classification, with a specificity of over 64%, MCC of over 76%, and balanced accuracy of 81%.

Table 10. Models' Accuracy Comparison.

	Accuracy
SVM	0.899
RF	0.988
NN	0.751
Bagging	0.977
Stacked Ensemble	0.9975

Table 11. Performance Metrics of Ensemble Model.

	Stacked Ensemble
ACC	0.9975
SENS	0.9979
SPEC	0.6418
MCC	0.7664
BA	0.8198

In summary, our study shows that hybrid sampling is an effective balancing technique for imbalanced financial datasets, and that stacking multiple high-performing individual models can lead to even better performance. To evaluate the performance of our proposed model, we compared it to existing literature, and previous studies reported accuracies ranging from 83% to 98% for similar ensemble stacking tasks [11,18,27]. These findings can inform the development of machine learning models for financial applications, and serve as a basis for future research in this field.

5 Conclusion

Financial bankruptcy is declared when a company can no longer to pay off its debts and obligations, and it can have significant economic consequences for both the company and its stakeholders. Researchers and financial institutions have thus sought methods that would improve the accuracy of its prediction, thereby mitigating the risk of financial losses. This study aimed to develop a novel machine learning model with a focus on achieving high accuracy in predicting bankruptcy. Remarkably, our proposed model demonstrated an outstanding accuracy of 99.75%. This exceptional performance highlights the effectiveness and reliability of our approach in accurately predicting bankruptcy. The substantial improvement achieved by our model, surpassing the accuracies reported in existing literature, underscores the advancement and contribution of our research. Based on the results obtained from this study, it can be concluded that the use of ensemble learning techniques in predicting financial bankruptcy can significantly improve the accuracy of predictions. The study also showed that the use of different sampling techniques can have a significant impact on the performance of models. In particular, the hybrid sampling technique used in this study proved to be effective in addressing the class imbalance problem commonly encountered in financial bankruptcy prediction. Overall, the findings of this study suggest that ensemble learning, coupled with appropriate sampling techniques, can be a powerful tool for predicting financial bankruptcy. Further research can be conducted to explore the effectiveness of weighted hard and soft voting as an ensemble learning technique on different types of datasets and to investigate the possibility of combining different types of models with the logit model in an ensemble.

CRediT Authorship Contribution Statement

Anthonia Oluchukwu Njoku: Conceptualization, Methodology, Validation, Formal analysis, Investigation, Software, Writing & editing, Visualization.
    Berthine Nyunga Mpinda: Validation, Formal analysis, Review & editing.
    Olushina Olawale Awe: Conceptualization, Methodology, Validation, Review & editing, Supervision, Project administration.

Declaration of Competing Interest

The authors declare that they have no known competing financial interests or personal relationships that could have appeared to influence the work reported in this paper.

Data Availability

The data and R codes used in this work can be found in: https://drive.google.com/drive/folders/1LdK_fdKicEf8qC_iYTcffF0UoRoBBwU5?usp=share_link

References

1. Adamus, R.: Bankruptcy proceedings in relation to bond issuers in Poland. Soc. Polit. Sci. **1**, 146–149 (2013)
2. Alam, T.M., et al.: Corporate bankruptcy prediction: an approach towards better corporate world. Comput. J. **64**(11), 1731–1746 (2021). https://doi.org/10.1093/comjnl/bxab095
3. Altman, E.I.: Financial ratios, discriminant analysis and the prediction of corporate bankruptcy. J. Financ. **23**(4), 589–609 (1968)
4. Altman, E.I.: Predicting financial distress of companies: revisiting the z-score and zeta$^{®}$ models. In: Handbook of Research Methods and Applications in Empirical Finance, pp. 428–456. Edward Elgar Publishing (2013)
5. Anjum, S.: Business bankruptcy prediction models: a significant study of the Altman's z-score model. Available at SSRN 2128475 (2012)
6. Balcaen, S., Ooghe, H.: 35 years of studies on business failure: an overview of the classic statistical methodologies and their related problems. Br. Account. Rev. **38**(1), 63–93 (2006). https://doi.org/10.1016/j.bar.2005.09.001
7. Barboza, F., Kimura, H., Altman, E.: Machine learning models and bankruptcy prediction. Expert Syst. Appl. **83**, 405–417 (2017). https://doi.org/10.1016/j.eswa.2017.04.006
8. Bertsimas, D., King, A.: Logistic regression: from art to science. Stat. Sci. **32**(3), 367–384 (2017)
9. Bondarenko, P.: Enron scandal. Encyclopedia Britannica (2019)
10. Butcher, B., Smith, B.J.: Feature Engineering and Selection: A Practical Approach for Predictive Models. Chapman & Hall/CRC Press, Boca Raton (2020)
11. Buyrukoglu, S., Savaş, S.: Stacked-based ensemble machine learning model for positioning footballer. Arab. J. Sci. Eng. **48**(3), 1371–1383 (2023)
12. Charan, R., Useem, J., Harrington, A.: Why companies fail. Fortune **27**, 36–44 (2002)
13. Domingos, P.: Bayesian averaging of classifiers and the overfitting problem. In: ICML, vol. 747, pp. 223–230 (2000)
14. Fauzi, M.A., Yuniarti, A.: Ensemble method for Indonesian twitter hate speech detection. Indones. J. Electr. Eng. Comput. Sci. **11**(1), 294–299 (2018)
15. Fitzpatrick, P.J.: A comparison of the ratios of successful industrial enterprises with those of failed companies. The Accountants' Magazine (1932)
16. Garcia, J.: Bankruptcy prediction using synthetic sampling. Mach. Learn. Appl. **9**, 100343 (2022). https://doi.org/10.1016/j.mlwa.2022.100343
17. Horak, J., Vrbka, J., Suler, P.: Support vector machine methods and artificial neural networks used for the development of bankruptcy prediction models and their comparison. J. Risk Financ. Manag. **13**(3), 60 (2020). https://doi.org/10.3390/jrfm13030060
18. Hosseini, S., Pourmirzaee, R., Armaghani, D.J., et al.: Prediction of ground vibration due to mine blasting in a surface lead-zinc mine using machine learning ensemble techniques. Sci. Rep. **13**, 6591 (2023)
19. Hołda, A.: Zasada kontynuacji działalności i prognozowanie upadłości w polskich realiach gospodarczych. Zeszyty Naukowe/Akademia Ekonomiczna w Krakowie. Seria Specjalna, Monografie (174) (2006)
20. Jones, S.: Corporate bankruptcy prediction: a high dimensional analysis. Rev. Acc. Stud. **22**, 1366–1422 (2017). https://doi.org/10.1007/s11142-017-9407-1

21. Khan, S.I., Hoque, A.S.M.L.: SICE: an improved missing data imputation technique. J. Big Data **7**, 37 (2020). https://doi.org/10.1186/s40537-020-00313-w
22. Kiaupaite-Grushniene, V.: Altman z-score model for bankruptcy forecasting of the listed Lithuanian agricultural companies. In: 5th International Conference on Accounting, Auditing, and Taxation (ICAAT 2016), pp. 222–234. Atlantis Press (2016)
23. Kim, M.J., Kang, D.K.: Ensemble with neural networks for bankruptcy prediction. Expert Syst. Appl. **37**(4), 3373–3379 (2010). https://doi.org/10.1016/j.eswa.2009. 10.012
24. Kitowski, J., Kowal-Pawul, A., Lichota, W.: Identifying symptoms of bankruptcy risk based on bankruptcy prediction models-a case study of Poland. Sustainability **14**(3), 1416 (2022)
25. Kou, G., et al.: Bankruptcy prediction for SMEs using transactional data and two-stage multiobjective feature selection. Decis. Support Syst. **140**, 113429 (2021). https://doi.org/10.1016/j.dss.2020.113429
26. Kufeoglu, S.: SDG-9: industry, innovation and infrastructure. In: Emerging Technologies (Sustainable Development Goals Series). Springer (2022). https://doi.org/10.1007/978-3-031-07127-0-11
27. Learning, S.M.: Hybrid model for twitter data sentiment analysis based on ensemble of dictionary-based classifier and stacked machine learning classifiers-SVM, KNN and c5.0. J. Theoret. Appl. Inf. Technol. **98**(04), 624–635 (2020)
28. Lian, W., Nie, G., Jia, B., et al.: An intrusion detection method based on decision tree-recursive feature elimination in ensemble learning. Math. Probl. Eng. **2020**, 1–15 (2020)
29. Lombardo, G., Pellegrino, M., Adosoglou, G., Cagnoni, S., Pardalos, P.M., Poggi, A.: Machine learning for bankruptcy prediction in the American stock market: dataset and benchmarks. Future Internet **14**(8), 244 (2022). https://doi.org/10.3390/fi14080244
30. Maddikonda, S.S.T., Matta, S.K.: Bankruptcy prediction: mining the Polish bankruptcy data (2018)
31. Malek, N.H.A., Yaacob, W.F.W., Wah, Y.B., et al.: Comparison of ensemble hybrid sampling with bagging and boosting machine learning approach for imbalanced data. Indones. J. Electr. Eng. Comput. Sci. **29**(1), 598–608 (2023)
32. Misankova, M., Bartosova, V.: Comparison of selected statistical methods for the prediction of bankruptcy. In: Conference Proceedings of 10th International Days of Statistics and Economics, Melandrium, Prague, pp. 895–899 (2016)
33. Muller, D., Soto-Rey, I., Kramer, F.: An analysis on ensemble learning optimized medical image classification with deep convolutional neural networks. IEEE Access **10**, 66467–66480 (2022)
34. Qu, Y., Quan, P., Lei, M., Shi, Y.: Review of bankruptcy prediction using machine learning and deep learning techniques. Procedia Comput. Sci. **162**, 895–899 (2019). https://doi.org/10.1016/j.procs.2019.12.065
35. Rodrigues, P.C., Awe, O.O., Pimentel, J.S., Mahmoudvand, R.: Modelling the behaviour of currency exchange rates with singular spectrum analysis and artificial neural networks. Stats **3**(2), 137–157 (2020)
36. Shekhar, S., Schrater, P.R., Vatsavai, R.R., Wu, W., Chawla, S.: Spatial contextual classification and prediction models for mining geospatial data. IEEE Trans. Multimedia **4**(2), 174–188 (2002)
37. Shetty, S., Musa, M., Brédart, X.: Bankruptcy prediction using machine learning techniques. J. Risk Financ. Manag. **15**(1), 35 (2022). https://doi.org/10.3390/jrfm15010035

38. Son, H., Hyun, C., Phan, D., Hwang, H.J.: Data analytic approach for bankruptcy prediction. Expert Syst. Appl. **138**, 112816 (2019). https://doi.org/10.1016/j.eswa.2019.06.050

39. Thilakarathna, C., Dawson, C., Edirisinghe, E.: Using financial ratios with artificial neural networks for bankruptcy prediction. In: 2022 IEEE International Conference on Artificial Intelligence and Computer Applications (ICAICA), Dalian, China, pp. 55–58. IEEE (2022). https://doi.org/10.1109/ICAICA54878.2022.9844640

40. Xu, Y., Klein, B., Li, G., Gopaluni, B.: Evaluation of logistic regression and support vector machine approaches for XRF based particle sorting for a copper ore. Miner. Eng. **192**, 108003 (2023)

Deep Learning and Machine Learning Techniques for Credit Scoring: A Review

Hana Demma Wube[1]([✉]) [iD], Sintayehu Zekarias Esubalew[1] [iD],
Firesew Fayiso Weldesellasie[1] [iD], and Taye Girma Debelee[1,2] [iD]

[1] Ethiopian Artificial Intelligence Institute, 40782 Addis Ababa, Ethiopia
hana.demma@gmail.com
[2] Electrical and Computer Engineering Department, Addis Ababa Science and
Technology University, Addis Ababa, Ethiopia

Abstract. Credit scoring is one of the most important credit decision-making in banking institutions by collecting, analyzing, and classifying various credit elements and variables of customer financial data. Nowadays, there is an increase in research related to machine learning (ML) and deep learning models (DL) to improve accuracy. This has led to the emergence of various ML and DL methods as a core practice in the field of credit scoring using various datasets. The aim of this study is to provide in-depth insights on various ML and DL-based credit scoring techniques. For this purpose, articles published between 2018 and 2023 were systematically reviewed by formulating research questions, defining search terms, and filtering articles using predefined inclusion and exclusion criteria. In particular, the reported model type, dataset, key performance parameters, publication profile, and keywords were extracted, and then the results of the identified models were examined. Finally, the most important aspects of the DL and ML methods in credit scoring were discussed. It was noticed that the performance of ML and DL-based credit scoring models has generally been evaluated using accuracy and area under curve. It was also observed that the UCI datasets have been used as a benchmark in the development of advanced credit scoring algorithms. The study also shows that comparing the performance of DL and ML models for credit is difficult due to the heterogeneity of the reported performance metrics. Hybrid and ensemble model based credit scoring techniques are becoming more popular and are the most commonly used credit scoring model. Further, the gaps and future research directions were highlighted. This review is expected to serve as an up-to-date and comprehensive reference for interested researchers seeking to quickly understand the current progress in DL and ML methods for credit scoring.

Keywords: Machine learning · Deep learning · Credit scoring ·
Performance parameter · Banking Institutions · Systematic literature
review · Dataset

T. G. Debelee et al. (Eds.): PanAfriConAI 2023, CCIS 2069, pp. 30–61, 2024.
https://doi.org/10.1007/978-3-031-57639-3_2

1   Introduction

Credit scoring is the process of collecting, analyzing, and classifying a variety of non-numerical and numerical customer financial data [33]. Using the outcomes of the credit scoring process, banking institutions can forecast the current and expected risk of bad creditworthiness of customers. The classification of good and bad bank loans is determined by variables such as educational level, time in the current job, loan amount, loan duration, monthly income, guarantees, purpose of loan, and others. In this regard, the ability to make the correct judgments between large credit datasets in the credit scoring process has become one of the most challenging tasks for banking institutions. Advances in computing power, as well as modern machine learning (ML) and deep learning (DL) techniques, simplify and speed up the process of making credit scoring decisions between large credit datasets. Various categories of credit scoring techniques, such as statistical techniques, ML, and DL techniques have been explored in previous studies [73,79,86].

Historically, statistical methods have long been used to build credit scoring models such as discriminant analysis and logistic regression that do not easily handle large datasets [62,79]. However, over the past decades, conventional ML techniques such as random forest (RFs) [16], support vector machines (SVMs) [18], decision trees (DTs) [42] and neural networks (NNs) [99] have become a more powerful, effective and flexible than statistical methods. In particular, in recent days, DL techniques such as convolutional neural networks (CNN) [65], long short-term memory networks (LSTM) [9] and deep belief network (DBN) [101] are widely used techniques for credit scoring that can provide the highest output accuracy. Further, hybrid methods (HYBRID) [47], and ensemble (COMBINED) models [74] such as boosting [75], stacking [69], bagging [32], have also been proposed to obtain a better accuracy. From the above literature, it can be understood that the development of efficient and sophisticated ML and DL methods has gained popularity and attracted researchers in the realm of credit scoring to make significant progress toward achieving higher accuracy of the credit decision process. Therefore, the current study aims to provide in-depth insights, into current progress, trends, challenges, and future research prospects on various ML and DL-based credit scoring methods.

This study is organized as follows. Section 2 presents related work on credit scoring using different methods. Section 3 furnishes an overview of the research methodology used in this study and outlines the research questions that will be addressed in the systematic literature review. Section 4 provides an analysis of selected existing literature and discusses key observations from the review, with particular emphasis on identifying ML and DL methods used for credit scoring, model type, dataset, key performance parameters, publication profile, potential challenges, and future prospects. Section 5 summarizes the overall findings of the study. The paper concludes by listing challenges and future work in credit scoring where the application of ML and DL can be further explored.

2   Related Works

As mentioned earlier, the use of ML and DL in the field of credit scoring is still an active research topic. In the current section selected review works that are directly related to the current study are presented in Table 1, despite the existence of several studies. The works were evaluated according to their contribution, similarity, and limitations. From the table, one can easily understand the need for up-to-date and comprehensive guidance on ML and DL models for credit scoring, which takes into account aspects of model type, data set, key per-

Table 1. A summary of related literature review works

Author	Contribution	Limitation
Markov et al. [59]	The study provides a systematic review of recent studies in the realm of credit scoring. Also, the study provided an overview of best practices and current trends in credit scoring, emphasizing those that have emerged only in the last few years.	–The relevant articles included in the study are only from the Science Direct database. –The study covers a limited set of ML methods, and more recent methods such as DL are not included.
Kumar et al. [46]	Provide an insight into how ML algorithms can be used to utilize credit scores in rural areas, The authors also provide an analysis of existing literature and suggest ways to enhance the accuracy of the prediction.	–The research work covers a limited set of ML methods, and more recent methods such as DL are not included.
Hayashi et al. [37]	–The research work provides a comprehensive overview of new trends in ML and DL methods for credit scoring. –The study presents a detailed analysis of the performance of the different algorithms with four datasets.	–The paper only analyzes studies published between 2019 and 2022. –The coverage of the database from which the included articles are retrieved is not specified and is not systematically selected. –The considered models are selected for a limited type of dataset.
Marques et al. [60]	The study provides a thorough review of the scientific articles that explore the application of evolutionary algorithms to credit scoring. Additionally, the paper discusses the key points of the review, highlighting the shortcomings and potential problems that should be addressed in the future.	–The study does not provide a comprehensive comparison of the different evolutionary algorithms used in credit scoring. –The study does not provide a thorough analysis of the performance of the different algorithms on different datasets. –The study considers articles published during the period 2000–2012.
Louzada et al. [57]	Conducted a systematic review and comparison of various classification methods for credit scoring	–The study did not include DL methods

formance, and publication profile. The current study can provide comprehensive information on the future prospects and potential of ML and DL methods in credit scoring for banking institutions and decision-makers.

3   Method

This section presents the methods utilized in the current systematic literature review to obtain sufficient data to discover a collection of research articles related to ML and DL techniques to support the credit scoring process.

In the current study, the following main steps are considered: (i) formulating research questions for review, (ii) defining search terms, (iii) filtering articles using predefined inclusion and exclusion criteria, (iv) capturing the most relevant and reliable studies based on the title, abstract, keywords and article full screening (v) extracting important information, (vi) summarizing and interpreting the results. The detailed description of the article search and selection process is shown in Fig. 1.

3.1   Research Questions (RQs)

A set of research questions has been developed to guide and achieve the objective of the current work. The five research questions along with their motivations are presented in Table 2.

Table 2. Research Questions

Research Questions	Aim to answer
RQ1: What are the ML and DL techniques for credit scoring that are most frequently used?	Identifying the most commonly used ML and DL techniques.
RQ2: What are the most recent study trends in the field of credit scoring?	Explore the latest research trends in credit scoring
RQ3: Types of dataset used for credit scoring	Identify the type of datasets used for credit scoring
RQ4: What are the key performance parameters and performance of the methods employed for credit scoring?	Identify the performance parameters and performance of the ML and DL-based credit scoring techniques.
RQ5: What are the challenges and limitations of using ML and DL techniques in credit scoring?	Identify the bottleneck of the credit scoring process while using ML and DL techniques
RQ6: What are the future research directions for improving credit scoring accuracy?	Identify and discuss the future direction of research

3.2   Article Search Strategy

To obtain the best possible results, specific search terms have been carefully chosen to ensure comprehensive coverage of relevant literature in six different databases according to the research questions. Article databases include Google

Scholar, ScienceDirect, IEEE Xplore, MDPI, SpringerLink, Wiley, and Emerald. The following specific search terms were used: (("machine learning" **OR** "deep learning" **OR** "artificial intelligence" **OR** "prediction") **AND** ("credit rating" **OR** "credit scoring" **OR** "loan evaluation" **OR** "credit bond" **OR** "creditworthiness" **OR** "credit risk" **OR** "credit risk assessment")). In this regard, the first set of terms ("machine learning," "deep learning," "artificial intelligence," and "prediction") were used to identify papers that discuss or utilize these techniques. The second set of terms ("credit rating," "credit scoring," "loan evaluation," "credit bond," "credit worthiness," "credit risk," and "credit risk assessment") was used to identify articles related to a particular topic of credit scoring. Logical operators were used to create search strings: "AND" was used to combine terms, while "OR" was used for synonyms. The above search terms were used in all six databases. The search was carried out between February and June 2023. A detailed procedure for defining search strings is depicted in Algorithm 1.

Algorithm 1. Pseudo-code for defining search strings

1: **Procedure** Title (DL and ML techniques in Credit Scoring)

2: **Search String** ← [(credit scoring" OR credit rating" OR loan evaluation" OR credit bond" OR credit worthiness" OR credit risk" OR Credit risk assessment") AND (Machine Learning Methods" OR Deep Learning Methods" OR Artificial intelligence") AND (Prediction" OR Classification")]

3.3 Inclusion and Exclusion Criteria

Inclusion and exclusion criteria were pre-defined to select relevant studies, as indicated in Table 3. A detailed description of article searching using inclusion and exclusion criteria is presented in Algorithm 2.

Table 3. Inclusion and Exclusion criteria

Inclusion Criteria	Exclusion Criteria
IQ1: The paper must be written in English	EQ1: The paper is not written in English.
IQ2: The paper must be published in a peer-reviewed journal or conference proceedings	EQ2: The paper is not published in a peer-reviewed journal or conference proceedings.
IQ3: The paper must discuss the use of ML or DL techniques in credit risk assessment, credit scoring, credit rating, loan evaluation, credit bond, credit worthiness, or credit risk assessment prediction.	EQ3: The paper lacks sufficient detail on the methods, techniques, and algorithms used to perform the analysis.
IQ4: The paper must be published between the years 2018 and 2023	EQ4: paper is published before the year 2018.
IQ5: The study that are relevant to the research question	EQ5: Studies that are not relevant to the research question

Algorithm 2. Pseudo-code for retrieving articles based on inclusion and exclusion criteria

```
1: Search_Databases = [Google Scholar, Springer Link, MDPI, Science Direct, Wiley Online,
   IEEE Xplore]
2: Area_keywords = ["credit scoring", "credit rating", "loan evaluation", "credit bond",
   "creditworthiness","credit risk", "credit risk assessment"]
3: Method_keywords = ["Deep Learning", "Machine Learning"]
4: Target_keywords = ["Classification", "Prediction"]
5: Search_String = "Algorithm 1"
6: Inclusion_Criteria = ["IC1", "IC2", "IC3", "IC4", "IC5"]
7: Exclusion_Criteria = ["EC1", "EC2", "EC3", "EC4", "EC5"]
          # Initialize the lists
8: scholarly_list1 = []
9: scholarly_list2 = []
10: scholarly_final_list = []
          # Search the databases
11: for keyword in Area_keywords do
12:     for target in Target_keywords do
13:         for method in Method_keywords do
14:             search_query ← Search_String + " " + keyword + " " + target + " " + method
15:             for database in Search_Databases do
16:                 paper_list = database.Search(search_query)
17:                 scholarly_list1.extend(paper_list)
18:             end for
19:         end for
20:     end for
21: end for
          # Apply the inclusion and exclusion criteria
22: for Paper in scholarly_list1 do
23:     if paper.meets_all(Inclusion_Criteria) and not paper.meets_any(Exclusion_Criteria)
    then
24:         scholarly_list2.append(paper)
25:     end if
26: end for
27: scholarly_final_list = apply_inclusion_criteria(scholarly_list2, Inclusion_Criteria,
    Exclusion_Criteria)
28: function APPLY_INCLUSION_CRITERIA(paper_list, inclusion_criteria, exclusion_criteria)
29:     final_list = []
30:     for paper in paper_list do
31:         if paper.meets_all(inclusion_criteria) and not paper.meets_any(exclusion_criteria)
    then
32:             final_list.append(paper)
33:         end if
34:     end for
35:     return final_list
36: end function
```

3.4 Article Filtering

After extracting the articles from the databases using search terms and inclusion criteria, the next step is article filtering. The article filtering process involves several steps, such as reviewing titles and abstracts, reading the full text, and evaluating the quality of the articles. Article filtering begins with screening out irrelevant articles by reading the titles and abstracts of articles. The screening process resulted in about 172 articles out of 722 articles. Subsequently, the authors read the full text of the selected articles and assessed their quality. At this stage, about 50 articles were filtered. A detailed article filtering procedure is presented in Algorithm 3 and flow chart on Fig. 1.

Algorithm 3. Pseudo-code for Article filtering

1: *search_results* ← function_to_search_database(*search_query*) ▷ returns a list of papers
2: **Input:** *search_query* - a string query to search the database
3: **Output:** *paper_list* - a list of papers matching the search query
4: *paper_list* ← function_to_search_database(*search_query*)
5: **return** *paper_list*

6: *final_paper_list* ← []
7: **for** *paper* in *search_results* **do**
8:   **if**   *paper.title_contains(search_query)*   **and** *paper.abstract_contains(search_query)* **then**
9:       *paper.read_full_paper()* ▷ this method reads the full paper and stores it in the paper object
10:       **if** *paper.full_paper_contains(search_query)* **then**
11:          *final_paper_list.append(paper)*
12:       **end if**
13:   **end if**
14: **end for**
15: **return** *final_paper_list*

---

4 Performance Measuring Metrics

The performance, accuracy, and predictive power of ML or DL credit scoring techniques can be measured using performance measuring metrics that include area under curve (AUC), accuracy (ACC), Recall (R), F1-score, and Precision (P). The performance measuring metrics are defined in terms of true positive (*TP*), false positive (*FP*), true negative (*TN*), and false negative (*FN*) values. *TP* and *TN* denote that the model correctly predicts the positive class and negative class, respectively FP and *FN* denote that the model incorrectly predicts the positive class and negative class, respectively. A brief description of the performance measuring parameters along with a mathematical description is presented in the following.

Accuracy (ACC): Measures the overall correctness of the model's predictions and is mathematically expressed as:

$$ACC = \frac{TP+TN}{TP+TN+FP+FN} \tag{1}$$

Precision (P): Measures the proportion of true positives among the total number of positive predictions and is mathematically defined as:

$$P = \frac{TP}{TP+FP} \tag{2}$$

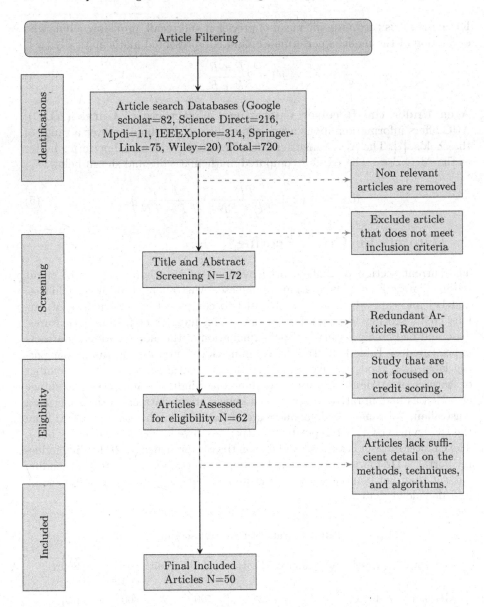

Fig. 1. Flow chart of search and selection of articles

Recall (R): measures the percentage of true positives among the total number of actual positive cases and is represented mathematically as:

$$R = \frac{TP}{TP + FN} \tag{3}$$

F1 Score: It is the harmonic mean of precision and recall, providing a balanced evaluation of the model's performance, and is represented mathematically as:

$$F1 = 2\frac{P \times R}{P + R} \tag{4}$$

Area Under the Receiver Operating Characteristics Curve (AUC): AUC offers information about a model's ability to discriminate over a range of thresholds. [7]. The AUC range is in the range [0, 1]. The model performs better as the AUC rises. The AUC is computed mathematically and shown below:

$$AUC = \frac{1}{2}\left(1 + \frac{TP}{TP + FN} - \frac{FP}{FP + TN}\right) \tag{5}$$

5   Datasets for Credit Scoring

The current section presents a brief overview of the characteristics of credit scoring datasets such as German, Japanese, Australian, Taiwanese, Chinese, and US mid-cap, which are mainly used to compare the performance of various credit scoring techniques in terms of the number of instances, features, good and bad credit history. A brief explanation of the aforementioned datasets is presented in Table 4. In Table 4, column "Good" indicates instances percentage labeled as good/positive samples, column "Features" indicates the number of features or attribute variables, column "Bad" indicates instances percentage classified as bad/negative samples. The known credit datasets such as German, Australian, Taiwanese, and Japanese credit datasets can be easily accessed from the UC Irvine (UCI) ML repository (https://archive.ics.uci.edu/ml/index.php). The Chinese dataset is mainly available on the comprehensive P2P lending industry portal WDZJ (www.wdzj.com). Lending Club (LC) is a peer-to-peer lending dataset that can be accessed from the website (https://www.lendingclub.com/info/download-data.action).

Table 4. Dataset for credit scoring

Dataset	Number of Instances	Features	Good credit	Bad credit	Source
German	1000	24	700	300	[23]
Japanese	690	15	307	383	[23]
Australian	690	14	307	383	[23]
Taiwanese	30,000	23	23,366	6634	[23]
Chinese	1057	10	552	505	[20]
Lending Club	133,887	90	23 999	109 888	[82]

6 Review of Recent ML and DL Techniques in Credit Scoring

This section briefly introduces the most popular computational methods used for credit scoring, such as ML and DL methods, each of which has similar principles and unique characteristics.

6.1 Machine Learning Techniques for Credit Scoring

ML is a popular approach to credit scoring owing to its potential to analyze large amounts of data and uncover complex patterns that traditional statistical methods can miss. Some of the ML techniques are briefly discussed below.

Decision Trees (DTs): DTs work by recursively splitting the input data into smaller subsets with the help of the most important feature. Each section is designed to reduce data uncertainty, resulting in a tree structure that can be easily visualized. DTs can handle both categorical and numerical data and are easy to interpret, making them the most popular in the credit scoring process. Further, Baesens et al. showed that DTs were among the most accurate ML algorithms for credit scoring [6]. Furthermore, Teles et al. pointed out that DTs can help to understand sequential decisions and outcome dependencies. DT model can also play a complementary role to other credit scoring tools such as fuzzy assets, and the classes it produces can be used as fuzzy sets. However, the DT algorithm requires a target attribute with only discrete values. Another drawback is that it does not handle complex interactions well. Additionally, DT is too sensitive to noise, irrelevant attributes, and the training data set, and DT is mostly a classification model in preference to a predictive model [81].

Random Forest (RF): RF is an ensemble learning technique that integrates several DTs to create a more accurate model and has the potential to handle numerical categorical data, which is important in credit scoring where many factors are considered. The RF method is widely used in credit scoring [11,80]. A study by Li et al. revealed that RF has better performance in credit scoring applications in terms of model operating time, ACC, AUC, Kolmogorov-Smirnov statistic, and Brier score [50].

Support Vector Machines (SVMs): SVMs is a competitive supervised binary classifier. SVMs can be used for inseparable and linearly separable data [83]. Goh and Lee reported that SVMs yield good performance in credit scoring [30]. Further, Bhatore et al. reported that SVM-based approaches can be applied to overcome local optimum and over-fitting issues in ANN-inspired models. Despite the advantages, the long training time and comprehensibility of these models are major challenges in applying SVM in credit scoring [14].

Neural Networks (NN): NN are useful in computing a target variable from various independent factors. NN comprises an input layer, multiple hidden layers in between, and an output layer. Additionally, the nodes are interconnected across these layers. NN consists of an input layer, a set of hidden layers in between, and an output layer with interconnected nodes. NN ability in credit scoring has been reported in the reference [5,8]. In addition, Bathore et al. reported that feed-forward neural networks are widely used among NN methods due to their generalizability property. However, the interpretability of these models is a serious drawback that makes it difficult for the person responsible for issuing loans to understand the process that the model follows [14].

Ensemble Model (COMBINATION). Ensemble learning is formed by combining the main classifiers. Examples of ensemble models are boosting, bagging, and stacking. Among the ensemble models, boosting is the most powerful model [92]. A new model in Bagging and Boosting is created by merging similar types of classification algorithms by assigning weights to the model and a weighted average in the case of a numerical classification problem. However, all models are assigned the same weight in bagging, whereas different weights are assigned in boosting. In addition, models are built separately in bagging, while models are built iteratively in boosting [92]. Stacking is another type of ensemble model where different algorithms can be combined. In Ensemble model, decision-making without assigning weights to the algorithm is possible as far as the classifiers have relatively the same performance, but if two of the three algorithms have poor performance, this can greatly affect the results [29]. Adaptive boosting ML algorithm (AdaBoost) is another type of boosting algorithm. AdaBoost uses a combination of a set of classifiers generated by the learning algorithm. The classifier models need to be the same type, and the model is formed with the help of instance weights [76]. The study reported by Koutanaei confirms that ensemble learning classification algorithms employed for credit scoring have superior performance than single classifiers [45]. Further, Wang et al. reported that Bagging performs better than Boosting [89]. Furthermore, Xia et al. demonstrated that stacking of heterogeneous credit-scoring ensemble models outperforms the benchmark individual homogeneous ensemble models [95]. Finally, Table 5, 6, 7, 8, 9 and 10 presents various ML methods for credit scoring obtained from the current systematic literature review in terms of dataset, model, and performance.

6.2   Deep Learning Techniques for Credit Scoring

DL is a series of models that can extract deep features from input data using a deep neural network (DNN) architecture. DL models most commonly have more than three layers. A DNN is initialized with unsupervised layer-by-layer learning and then tuned up with supervised learning with labels that can incrementally generate layer-by-layer higher-level features [13]. Research has shown that DL approaches outperform traditional statistical and classical ML models in credit scoring due to their ability to extract meaningful patterns from large and complex datasets. [36]. However, there are reports that DL has not been widely adopted for credit scoring due to data imbalance and scale, which influences the model performance [63]. Currently, Several DL models for credit have been applied and reported in the literature such as LSTM networks, Convolutional neural networks, Autoencoders, and ensemble systems [79]. The most

Table 5. Summary of ML techniques on credit scoring for the German dataset

Author	Method	ACC (%)	AUC	F1 (%)	P (%)	R (%)
Jadhav et al. [40]	SVM	82.8	–	–	–	–
Tripathi et al. [84]	Evolutionary Extreme Learning Machine	76.31	0.8123	–	–	–
Liu et al. [53]	Multi-grained augmented gradient boosting decision trees (mg-GBDT)	77.15	0.7929	–	78.47	91.86
Zhang et al. [103]	Ensemble of LR+SVM+RF+GBDT+NN and CF-Ens	76.15	0.7955	84.20	–	–
Pandey et al. [68]	RF with recursive elimination of features attribute	83.3	0.909	–	–	–
Yotsawat et al. [100]	Cost-Sensitive Neural Network Ensemble (CS-NNE)	74.40	0.8011	–	–	–
Goh et al. [31]	AI-Hybrid-RF	75.60	0.8053	84.10	–	–
Bastos [12]	Boosted decision trees (BDT)	–	0.811	–	–	–
Almustfa [39]	RF-Rand over sampling	85.4	0.925	–	–	–
Guo et al. [35]	GBDT	78.3	0.806	–	–	–
Dastile et al. [21]	RF	74	0.76	–	–	–
Alasbahi and Zheng [3]	Extreme Learning Machine	71.10	0.64	64.80	65.20	64.50
Boughaci and Alkhawaldeh [15]	Variable neighborhood search method (VNS)+SVM	77.46	–	–	–	–
Liu and Pan [55]	KNN+SVM	98.60	–	–	–	–
Tripathi et al. [85]	Radial Basis Function Neural Network (RBFN)	84.38	–	–	–	–
Shen et al. [78]	AdaBoost Back-Propagation Neural Network + AdaBoost	83	–	–	–	–
Xu et al. [98]	Ensemble of Extreme Learning Machine and Generalized Fuzzy Soft Sets	76.2	0.764	–	–	–
Liu et al. [52]	AugBoost-RFS	93.34	0.9756	–	84.58	78.98
Kazemi et al. [43]	GA+NN	87.1	–	–	–	–

common DL techniques used for credit scoring are depicted in Fig. 2. Further, in the following, some of DL methods are briefly discussed below.

Shallow Neural Network (SNN): SNN is an ANN with at most one hidden layer. SNN includes a Convolutional Neural Network with one hidden layer (CNN1), a Multilayer Perceptron with one hidden layer (MLP1), a radial basis function network (RBF), and a Single-layer Perceptron (SLP). CNN1 is sometimes also called one-dimensional convolution neural networks (1D-CNNs). In Alasbahi and Zheng study, SNN model was used for automated credit scoring [3].

Table 6. Summary of ML techniques for credit scoring for an Australian dataset

Author	Method	ACC (%)	AUC	F1 (%)	P (%)	R (%)
Jadhav et al. [40]	SVM	90.7514	–	–	–	–
Tripathi et al. [84]	Evolutionary Extreme Learning Machine	86.98	0.9250	–	–	–
Liu et al. [53]	Multi-grained augmented gradient boosting decision trees (Mg-GBDT)	76.31	0.8123	–	–	–
Zhang et al. [103]	Ensemble of LR+SVM+ RF+GBDT+NN and CF-Ens	87.17	0.9312	86.17	–	–
Dumitrescu et al. [25]	SVM	–	0.9210	–	–	–
Yotsawat et al. [100]	Cost-Sensitive Neural Network Ensemble (CS-NNE)	84.93	0.9131	–	–	–
Goh et al. [31]	AI-Hybrid-RF	87.38	0.9366	86.14	–	–
Bastos [12]	Boosted decision trees (BDT)	–	0.94	–	–	–
Wei et al. [91]	Noise-Adapted 2 Layer Ensemble Model	87.923	0.95584	–	–	–
Boughaci and Alkhawaldeh [15]	Variable neighborhood search method (VNS) +SVM	86.50	–	–	–	–
Liu and Pan [55]	KNN+SVM	91.87	–	–	–	–
Tripathi et al. [85]	RBFN	88.31	–	–	–	–
Shen et al. [78]	AdaBoost Back-Propagation Neural Network+ AdaBoost	83	–	–	–	–
Xu et al. [98]	Ensemble of Extreme Learning Machine and Generalized Fuzzy Soft Sets	73.5	0.852	–	–	–
Kazemi et al. [43]	GA+NN	97.78	–	–	–	–

Table 7. Summary of ML techniques for credit scoring for a Japanese dataset

Author	Method	ACC (%)	AUC	F1 (%)	P (%)	R(%)
Tripathi et al. [84]	Evolutionary Extreme Learning Machine	83.09	0.8930	–	–	–
Liu et al. [53]	Multi-grained augmented gradient boosting decision trees (Mg-GBDT)	87.08	0.9366	–	88.06	88.82
Zhang et al. [103]	Ensemble of LR+SVM+RF+ GBDT+NN and CF-Ens	86.96	0.9328	88.00	–	–
Wei et al. [91]	Noise-Adapted 2-Layer Ensemble Model	89.130	0.95069	–	–	–
Tripathi et al. [85]	RBFN	87.34	–	–	–	–
Xu et al. [98]	Ensemble of Extreme Learning Machine and Generalized Fuzzy Soft Sets	75.1	0.882	–	–	–

Table 8. Summary of ML techniques for credit scoring for a Lending Club dataset

Author	Method	ACC (%)	AUC	F1 (%)	P (%)	R (%)
Luca Zanin [102]	Extreme Gradient Boosting (XGBoost)	68.42	0.6886	32.71	23.39	54.37
Yotsawat et al. [100]	Cost-Sensitive Neural Network Ensemble (CS-NNE)	63.61	0.7082	–	–	–
Goh et al. [31]	AI hybrid-RF	85.72	0.8679	90.64	–	–
Chang et al. [17]	2-layer NN	87.06	–	87.49	84.61	90.56
Xu et al. [98]	Ensemble of Extreme Learning Machine and Generalized Fuzzy Soft Sets	71.2	0.671	–	–	–

Table 9. Summary of ML techniques for credit scoring for a Chinese dataset

Author	Method	ACC(%)	AUC	F1 (%)	P (%)	R (%)
Zhang et al. [104]	Cost-sensitive multiple-instance learning method (MIL)	83.09	0.89	91	100	84
Zhu et al. [107]	Esembleof RF,XGBoost, LightGBM and GBDT	99.02	–	–	–	–
Niu et al. [66]	LightGBM	66.22	0.711	65.9	–	–

Table 10. Summary of ML techniques for credit scoring for a Taiwanese dataset

Author	Method	ACC (%)	AUC	F1 (%)	P (%)	R (%)
Jadhav et al. [40]	SVM	82.5733	–	–	–	–
Liu et al. [53]	Mg-GBDT	69.63	0.7503	–	73.56	61.37
Dumitrescu et al. [25]	NN	–	0.7304	–	–	–
Munkhdalai et al. [64]	Multi-Layer Perceptron (MLP)	81.9	0.754	–	–	–

Deep Neural Networks (DNNs): DNNs is an ANNs with multiple hidden layers between the input and the output layers. DNNs have been widely used for credit risk prediction, particularly in predicting loan default [51]. In a study by Lin et al., a DNN was used to predict loan default using a Chinese peer-to-peer dataset [51]. The popular variants of DNN include Recursive Neural Networks (RvNN), Stacked Auto-encoders (SAE), Deep Belief Networks (DBN), Convolutional Neural Networks with multiple hidden layers (CNNm), Recurrent Neural Networks (RNN) and Multilayer Perceptron with multiple hidden layers (MLPm) [24]. Some of the variants of DNN are described below.

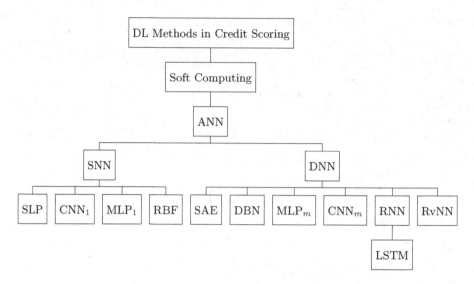

Fig. 2. DL techniques used in credit scoring

Recurrent Neural Networks (RNNs): RNNs have been used to model sequential data to capture temporal patterns in credit behavior [10]. A study by Livieris reported that trained RNNs can provide a better classification efficiency compared to a single classifier in a credit scoring application [56]

Long Short-Term Memory (LSTM): LSTM is one of the RNN types in which feedback links are connected with some layers of the network. LSTM is more suitable for learning from experience in time series forecasting with arbitrary size time steps, unlike conventional RNNs. In addition, LSTM solves the vanishing gradient problem using memory block store time-related information method for an arbitrary amount of time [38]. Moreover, Adisa et al. demonstrated that an optimized LSTM model shows a superior prediction performance than a conventional LSTM model and the ensemble models [2]. Further, the LSTM model can be applied to an online financial credit scoring [22].

Autoencoders (AEs): Autoencoders (AEs) are ANNs in which the input and output coincide. AEs operate by first compressing the input into a latent-space representation, which is then used to reconstruct the output. AEs consist of two main components such as encoder, which accepts input and compresses it into a smaller form, and a decoder, which attempts to recover the input. Among AEs, Stacked Autoencoders (SAEs) are DNNs, whereby each hidden layer is connected to the input of the next hidden layer [71]. AEs have been used for credit scoring to extract meaningful features from multivariate credit data [96]. Further, SAEs are used to overcome the adverse effect of imbalanced data in credit scoring problems [93].

Deep Belief Networks (DBNs): DBNs is another variant of ANN that is composed of a stack of Restricted Boltzmann Machines (RBMs). DBNs are "restricted" to one hidden layer and one visible layer, where connections are formed between the layers [61]. DBNs have been used to model complex relationships between credit characteristics and enhance credit scoring model accuracy [67]. Further, Liu et al. reported that DBN-based credit scoring model classification performance is superior to SVM and MLP models [58].

Generative Adversarial Networks (GANs): A typical GAN consists of two NNs competing with each other. GANs are currently being applied for oversampling minority class approaches to address class imbalance and improve credit scoring performance. [26,48].

Finally, Table 11, 12, 13, 14, 15, 16 and 17 presents various DL methods for credit scoring obtained from the current systematic literature review in terms of dataset, model, and performance.

Table 11. Summary of DL techniques on credit scoring for Japanese dataset

Author	Method	ACC (%)	AUC
Liu et al. [54]	Multi-layered gradient boosting DT	87.62	0.935
Li et al. [49]	Discriminant model based on Deep Forest	89.96	0.962
Du and Shu [22]	BRNN + LR + XGBoost	88.76	0.9491
Wang et al. [90]	Balanced incremental deep Q-network based on variational autoencoder data augmentation (BIDQN-VADA)	97.23	0.9727
Xu et al. [97]	DBN with AEnet-based feature selection	76.5	0.918

Table 12. Summary of DL based techniques on credit scoring for US dataset

Author	Method	ACC (%)	AUC
Korangi et al. [44]	LSTM	87.62	0.935
Korangi et al. [44]	TCN	89.96	0.962

Table 13. Summary of DL techniques on credit scoring for Australian dataset

Author	Method	ACC (%)	AUC
Plawiak et al. [70]	Deep genetic cascade ensemble of SVM	97.39	–
Gunnarsson et al. [34]	DBN	–	1.0
Liu et al. [54]	Multi-layered gradient boosting decision tree	88.2	0.9407
Dastil and Celik [19]	2D Convolutional Neural Networks (CNNs)	95	0.96
Du and Shu [22]	BRNN + LR + XGBoost	89.93	0.9574
Wang et al. [90]	BIDQN-VADA	95.00	0.9580
Xu et al. [97]	DBN with AEnet-based feature selection	75.8	0.916

Table 14. Summary of DL based techniques on credit scoring for Lending Club (LC) dataset

Author	Method	ACC (%)	AUC
Zhong and Wang [106]	Random under-sampling-Deep forest model (RUS-DF)	–	0.9968
Zhang et al. [105]	Gradient boosting decision tree (GBDT) + NN	–	0.7208
Liu et al. [54]	Multi-layered gradient boosting DT	67.86	0.7374
Abdoli et al. [1]	Bagging Supervised Autoencoder Classifier (BSAC)	–	0.9353
Qian et al. [72]	End-to-end soft reordering one-dimensional CNN (SR-1D-CNN)	–	0.7271
Wang et al. [90]	BIDQN-VADA	99.57	0.9913

Table 15. Summary of DL based techniques on credit scoring for German dataset

Author	Method	ACC (%)	AUC
Shen et al. [77]	Ensemble of LSTM and Ada Boost	–	0.8032
Gunnarsson et al. [34]	DBN	–	1.0
Liu et al. [54]	Multi-layered gradient boosting DT	76.53	0.7846
Li et al. [49]	Discriminant model based on Deep Forest	89.96	0.962
Jiao et al. [41]	CNN-XGBoost	84.72	–
Dastil and Celik [19]	2D Convolutional Neural Networks (CNNs)	88	0.80
Du and Shu [22]	BRNN + LR + XGBoost	77.50	0.8374
Veeramanikandan and Jeyakarthic [87]	Deep Neural Network (DNN)+Stacked Autoencoders (SA) + Truncated Backpropagation Through Time (TBPTT)	96.10	–
Fanai and Abbasimehr [27]	AE-DNN	–	0.5673
Wang et al. [90]	BIDQN-VADA	97.99	0.9761
Xu et al. [97]	DBN with AEnet-based feature selection	75.1	0.761

Table 16. Summary of DL-based techniques on credit scoring for Taiwanese dataset

Author	Method	ACC (%)	AUC
Shen et al. [77]	Ensemble of LSTM and Ada Boost	–	0.8032
Gunnarsson et al. [34]	DBN	–	1.0
Dastile et al. [19]	2D-CNN	88	–
Liu et al. [54]	Multi-layered gradient boosting decision tree	69.22	0.7490
Li et al. [49]	Discriminant model based on Deep Forest	89.96	0.962
Lei et al. [48]	Imbalanced generative adversarial fusion network (IGAFN)	85.65	0.7357
Abdoli et al. [1]	Bagging Supervised Autoencoder Classifier (BSAC)	–	1.3481
Ala'raj et al. [4]	MP-LSTM	82.03	0.78
Wang et al. [90]	BIDQN-VADA	84.50	0.7238

Table 17. Summary of DL-based techniques on credit scoring for Chinese dataset

Author	Method	ACC (%)	AUC
Wu et al. [94]	Deep multiple kernel classifier (DMKC)	87.62	0.935
Wang et al. [88]	Attention Mechanism LSTM (AMLSTM)	–	0.669
Ding and Yang [88]	DNN	–	0.7670
Feng et al. [28]	CCR-CNN	95.25	–

7 Discussion

This section presents the results of a comprehensive systematic literature review of selected publications according to the RQs.

7.1 Study Characteristics

The current section presents the study characteristics of the included articles such as the year of publication, geographical trends, co-occurrence of keywords and publication profiling.

7.1.1 Year of Publication

The publication years of the 50 included studies ranged from 2018 to 2023. Figure 3 illustrates the frequency distribution of publications growth from 2018 through 2023. It can be observed from the figure that 2021 attained a higher annual production rate of 65.21%.

7.1.2 Geographical Trends

In total, 17 countries participated in the total number of publications. Figure 4 presents the research contributions of various countries in the realm of DL and ML-based credit scoring techniques according to several published studies. Although researchers around the world are studying ML and DL techniques for credit scoring, most of the studies are mainly conducted in the People's Republic of China (23 studies), India (5 studies), Iran (3 studies), and The United Kingdom (3 studies). The remaining studies are distributed among Europe (6 studies), other Asian countries (6 studies), Africa (3 studies), and the Arab world (1 study). According to the aforementioned observations, It can be argued that a significant amount of the research originates from China.

7.1.3 Co-occurrence of Keywords

The co-occurrence of keywords analysis was performed using VOSviewer and the network graph of the keywords is presented in Fig. 5. In the figure, the node size indicates the frequency of occurrence. Curves between nodes show their overlap in one publication. The shorter the distance between two nodes, the more matches between two keywords. There are 157 different keywords provided by the authors. As can be noticed from Fig. 5, the most frequently used terms are

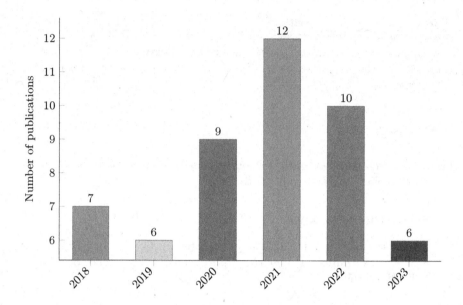

Fig. 3. Year of Publication

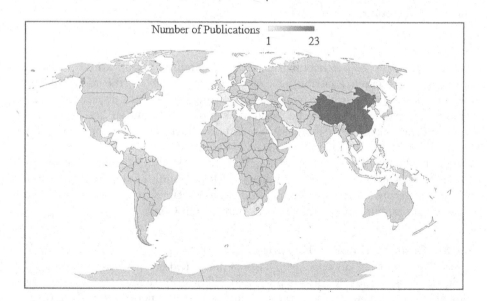

Fig. 4. Distribution of included studies by country based on corresponding authors

located in the center of the graph and have the strongest linkage with other key-words. The most frequently used terms are "credit scoring", "machine learning", "deep learning", and "feature selection", with 32 occurrences, 16 occurrences, 9 occurrences, and 8 occurrences, respectively. Moreover, the term "classification" and "ensemble model" appears 5 times.

From the keyword analysis, it can be observed that the ML techniques are frequently studied compared to DL techniques in credit scoring, which can also be observed in Fig. 6. Further, it can also be observed that the recent research in the field of credit scoring is focused on feature selection than the of classification model. Furthermore, it can be noticed that the ensemble models are the most studied topics.

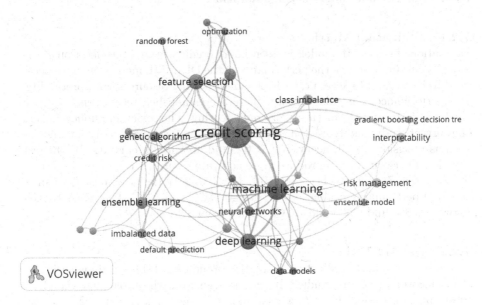

Fig. 5. Keywords network of publications

7.1.4 Publications' Profiling

In the current study, most of the 50 publications retrieved were journal articles (92%) and the remaining 8% were conference proceedings articles. This indicates the maturity of the studies. Further, the publications included in the current review have been indexed in both Web of Science and Scopus, showing that the review covers the most influential and accessible articles of the scientific community on ML and DL-based credit scoring methods. The breakdown of publications by their type is shown in Table 18.

Table 18. Types of articles.

Type of publication	No. of publications
Conference Proceeding	4 (8%)
Journal Article	46 (92%)
Total	50

7.2 Content Analysis

This section highlights the content of the included articles in terms of evaluation metrics, dataset usage, and model performance.

7.2.1 Evaluation Metrics

Evaluation of DL or ML models is essential for credit scoring Consideration of the model's output. To check the performance of the DL or ML models, many aspects need to be taken into account, such as computational resources, interpretability, and performance. For this purpose, there are established recommendations in the field of ML and DL methods for credit scoring. The reviewed publications in the current study mainly use four types of model performance evaluation metrics such as precision, recall, accuracy (ACC), and area under curve (AUC) to test the model performance to produce more accurate results. It can be noticed from Table 4, 5, 6, 7, 8, 9, 10, 11, 12, 13, 14, 15, 16 and 17 that the most frequently reported performance evaluation metrics are ACC and AUC. The AUCs ranged between 0.868 and 0.909 and ACCs ranged from 82% and 97%.

7.2.2 Dataset Usage

A dataset can reflect credit data evaluation standards. Table 19 summarizes the dataset coverage of the studies. It can be seen from the table that the three most commonly used datasets are German (31.25%), Australian (22.9%), and Taiwanese (13.5%). This shows that the use of the dataset in the credit scoring application is particularly focused on public UCI datasets. This may be due to the confidentiality issue in the application domain.

In addition, the UCI datasets provided more information that can be used to build credit scoring models. Moreover, UCI datasets have more additional data that can be utilized to model ML and DL-based credit scoring techniques. Further, other types of datasets such as Lending Club (LC) and Peer-to-Peer (P2P) online lending platforms are also used to test the robustness of the models. From the results, it can be concluded that the UCI datasets were used as a reference due to their common use in the literature.

Table 19. Dataset coverage of the studies

Dataset	ML	DL	Total
Australian	15	7	22
German	19	11	30
Taiwanese	4	9	13
Japanese	6	5	11
Chinese	3	4	7
Lending Club (LC)	5	6	11
US	–	2	2

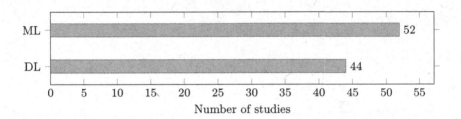

Fig. 6. Number of ML and DL studies

7.2.3 Model Performance To explore the performance relationships of the included study models and compare them with each other, several performance models are presented below. In terms of the ACC and AUC metric that measured the overall models' efficiencies, Fig. 7 depicts a scatter plot of studies that reported ACC and AUC values for the ML models (x and y axis, respectively) for various datasets. A total of 29 studies is included in the graph, such as [3, 21, 31, 35, 39, 52, 53, 64, 66, 68, 84, 91, 98, 100, 102–104] from ML studies and [4, 19, 22, 22, 44, 48, 49, 54, 54, 90, 94, 97] from DL studies were included. The studies closer to the upper right corner are the best, as they scored high values for both evaluation metrics. It can be seen that not all ML models show up in the scatterplot because not all studies report both metrics. It can also be noticed from the scatterplot that the model proposed by Wei et al. has the highest value of model performance (AUC = 0.95584, ACC = 87.923%) using Japanese dataset and (AUC = 0.95069, ACC = 89.130%) using Australian dataset as compared to other models [91]. Wei et al.'s best credit scoring model performance is obtained with a 2-layer noise-adapted isolated forest ensemble model based on a backflow learning approach [91].

Similarly, Fig. 8 depicts a scatterplot of studies that reported ACC and AUC values for the DL models (x and y axis, respectively) for various datasets. It can be seen from Fig. 8 that the model proposed by Wang et al. [90] has the highest model performance value (AUC = 0.9913, ACC = 99.57%) using Lending Club dataset and (AUC = 0.9580, ACC = 95.0%) using Australian dataset as compared to other models. Wang et al. [90]'s best credit scoring model perfor-

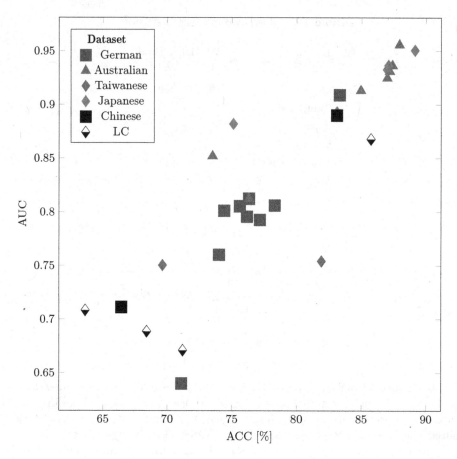

Fig. 7. The AUC and ACC scatterplot of ML models for credit scoring with different datasets.

mance is obtained by BIDQN-VADA. Also, it has been reported that the model proposed by Wang et al. has the advantage to deal with large datasets. In general, from Fig. 7 and Fig. 8, it can be concluded that the type of data set affects the ACC and AUC of the models, regardless of the models chosen, which are heterogeneous across studies. Further, it can also be concluded that comparing the performance of models is difficult due to the heterogeneity of the reported performance metrics.

Further, to identify the appropriate ML and DL models for credit scoring, it is reasonable to compare alternative models with varying datasets. In this regard, the models with the highest values of ACC and AUC for ML models and DL models are presented in Fig. 9 and Fig. 10, respectively. As shown in Fig. 9 the machine learning ensemble model for credit scoring has the highest accuracy

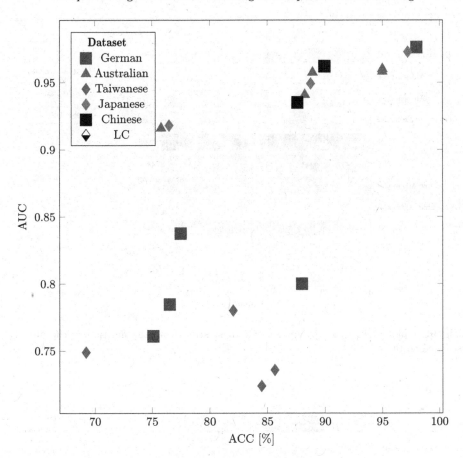

Fig. 8. The AUC and ACC scatterplot of DL models for credit scoring with different datasets.

(99.02%) for the Chinese dataset. It can also be observed from the figure that 54.55% (6 studies) achieved more than 97% accuracy. In addition, Fig. 10 shows the credit scoring methods with the highest AUC value. It can be noticed from Fig. 10 that the DBN model achieved the highest AUC value for Australian and German datasets compared to the others. It can also be noticed that more than 90% of the models achieved more than 0.9 AUC value. Overall, it can be concluded that hybrid models and ensemble models are growing in popularity and are most frequently applied in the field of credit scoring. Further, it can also be concluded that hybrid models and ensemble models are worthy of further study in credit scoring.

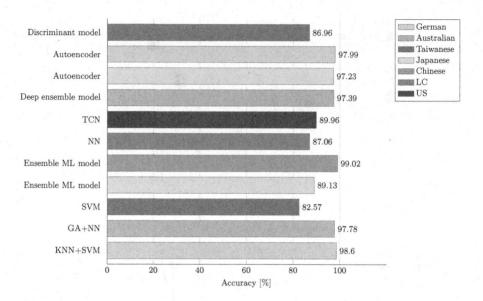

Fig. 9. A plot of accuracy by model. The best score from the datasets is the model selected to the graph.

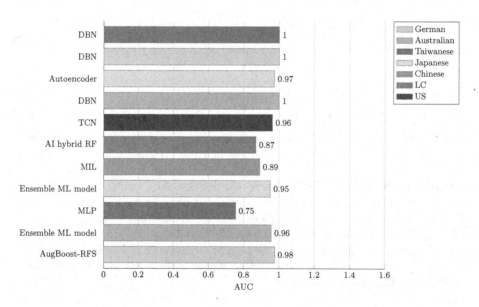

Fig. 10. A plot of AUC by model. The best score from the datasets is the model selected for the graph.

8   Conclusion and Future Work

This article focuses on a systematic literature review of ML and DL techniques for the advanced development of credit-scoring algorithms. The review process includes defining research questions, search strings, selecting relevant publication databases, and retrieving articles based on inclusion and exclusion criteria. Finally, after a critical evaluation of 50 publications published between 2018 and 2023, refined from a collection of 722 articles. The study shows that the performance of ML and DL-based credit scoring models has generally been evaluated using ACC and AUC. The leading research work on the application of ML and DL techniques for credit scoring is mainly performed in the People's Republic of China. The study also shows that ML techniques are frequently studied compared to DL techniques in credit scoring. The recent research in ML and DL-based credit scoring models are focused on feature selection rather than classification. The UCI datasets have been used as a benchmark in the development of advanced credit scoring algorithms. The study shows that comparing the performance of DL and ML models for credit is difficult due to the heterogeneity of the reported performance metrics. Hybrid and ensemble model-based credit scoring techniques are becoming more popular and are the most commonly used credit scoring models.

Further, to improve the quality of ML and DL research in the field of credit scoring in the future, the following recommendations are proposed. Feature selection methods are not discussed in the current study, however, feature selection methods are an important credit-scoring task for addressing the curse of dimensionality. Thus, feature selection methods may be a future research trend. Further study of hybrid and ensemble models, as well as incorporating private datasets can significantly contribute to the improvement and progress of credit scoring algorithms. Also, deep learning algorithms provide highly accurate predictions; however, they are often more difficult to understand, requiring in-depth study to make them easier to understand.

References

1. Abdoli, M., Akbari, M., Shahrabi, J.: Bagging supervised autoencoder classifier for credit scoring. Expert Syst. Appl. **213**, 118991 (2023)
2. Adisa, J., Ojo, S., Owolawi, P., Pretorius, A., Ojo, S.O.: Credit score prediction using genetic algorithm-LSTM technique. In: 2022 Conference on Information Communications Technology and Society (ICTAS), pp. 1–6. IEEE (2022)
3. Alasbahi, R., Zheng, X.: An online transfer learning framework with extreme learning machine for automated credit scoring. IEEE Access **10**, 46697–46716 (2022)
4. Ala'raj, M., Abbod, M.F., Majdalawieh, M., Jum'a, L.: A deep learning model for behavioural credit scoring in banks. Neural Comput. Appl. **34**, 1–28 (2022)

5. Angelini, E., Di Tollo, G., Roli, A.: A neural network approach for credit risk evaluation. Q. Rev. Econ. Finance **48**(4), 733–755 (2008)
6. Baesens, B.: Developing intelligent systems for credit scoring using machine learning techniques. Ph.D. thesis, Faculteit Economische en Toegepaste Economische Wetebnschappen, Katholieke Universiteit, Leuven (2003)
7. Baesens, B., Van Gestel, T., Viaene, S., Stepanova, M., Suykens, J., Vanthienen, J.: Benchmarking state-of-the-art classification algorithms for credit scoring. J. Oper. Res. Soc. **54**, 627–635 (2003)
8. Bahrammirzaee, A.: A comparative survey of artificial intelligence applications in finance: artificial neural networks, expert system and hybrid intelligent systems. Neural Comput. Appl. **19**(8), 1165–1195 (2010)
9. Bansal, M., Goyal, A., Choudhary, A.: A comparative analysis of k-nearest neighbour, genetic, support vector machine, decision tree, and long short term memory algorithms in machine learning. Decis. Anal. J. **38**, 100071 (2022)
10. Bao, W., Yue, J., Rao, Y.: A deep learning framework for financial time series using stacked autoencoders and long-short term memory. PLoS ONE **12**(7), e0180944 (2017)
11. Barboza, F., Kimura, H., Altman, E.: Machine learning models and bankruptcy prediction. Expert Syst. Appl. **83**, 405–417 (2017)
12. Bastos, J.A.: Predicting credit scores with boosted decision trees. Forecasting **4**(4), 925–935 (2022)
13. Bengio, Y., Lamblin, P., Popovici, D., Larochelle, H.: Greedy layer-wise training of deep networks. In: Schölkopf, B., Platt, J., Hoffman, T. (eds.) Advances in Neural Information Processing Systems, vol. 19. MIT Press (2006). https://proceedings.neurips.cc/paper_files/paper/2006/file/5da713a690c067105aeb2fae32403405-Paper.pdf
14. Bhatore, S., Mohan, L., Reddy, Y.R.: Machine learning techniques for credit risk evaluation: a systematic literature review. J. Bank. Financ. Technol. **4**, 111–138 (2020)
15. Boughaci, D., Alkhawaldeh, A.A.S.: Three local search-based methods for feature selection in credit scoring. Vietnam. J. Comput. Sci. **5**, 107–121 (2018)
16. Breiman, L.: Random forests. Mach. Learn. **45**, 5–32 (2001)
17. Chang, A.H., Yang, L.K., Tsaih, R.H., Lin, S.K.: Machine learning and artificial neural networks to construct P2P lending credit-scoring model: a case using lending club data. Quant. Finance Econ. **6**(2), 303–325 (2022)
18. Cortes, C., Vapnik, V.: Support-vector networks. Mach. Learn. **20**, 273–297 (1995)
19. Dastile, X., Celik, T.: Making deep learning-based predictions for credit scoring explainable. IEEE Access **9**, 50426–50440 (2021)
20. Dastile, X., Celik, T., Potsane, M.: Statistical and machine learning models in credit scoring: a systematic literature survey. Appl. Soft Comput. **91**, 106263 (2020). https://doi.org/10.1016/j.asoc.2020.106263. https://www.sciencedirect.com/science/article/pii/S1568494620302039
21. Dastile, X., Celik, T., Vandierendonck, H.: Model-agnostic counterfactual explanations in credit scoring. IEEE Access **10**, 69543–69554 (2022)
22. Du, P., Shu, H.: Exploration of financial market credit scoring and risk management and prediction using deep learning and bionic algorithm. J. Glob. Inf. Manag. (JGIM) **30**(9), 1–29 (2022)
23. Dua, D., Graff, C., et al.: UCI machine learning repository **7**(1) (2017). http://archiveics.uci.edu/ml
24. Duan, J.: Financial system modeling using deep neural networks (DNNs) for effective risk assessment and prediction. J. Franklin Inst. **356**(8), 4716–4731 (2019)

25. Dumitrescu, E., Hué, S., Hurlin, C., Tokpavi, S.: Machine learning for credit scoring: improving logistic regression with non-linear decision-tree effects. Eur. J. Oper. Res. **297**(3), 1178–1192 (2022)
26. Edla, D.R., Tripathi, D., Cheruku, R., Kuppili, V.: An efficient multi-layer ensemble framework with BPSOGSA-based feature selection for credit scoring data analysis. Arab. J. Sci. Eng. **43**(12), 6909–6928 (2018)
27. Fanai, H., Abbasimehr, H.: A novel combined approach based on deep autoencoder and deep classifiers for credit card fraud detection. Expert Syst. Appl. **217**, 119562 (2023)
28. Feng, B., Xue, W., Xue, B., Liu, Z.: Every corporation owns its image: corporate credit ratings via convolutional neural networks. In: 2020 IEEE 6th International Conference on Computer and Communications (ICCC), pp. 1578–1583. IEEE (2020)
29. Gicić, A., Subasi, A.: Credit scoring for a microcredit data set using the synthetic minority oversampling technique and ensemble classifiers. Expert. Syst. **36**(2), e12363 (2019)
30. Goh, R.Y., Lee, L.S.: Credit scoring: a review on support vector machines and metaheuristic approaches. Adv. Oper. Res. **2019** (2019)
31. Goh, R.Y., Lee, L.S., Seow, H.V., Gopal, K.: Hybrid harmony search-artificial intelligence models in credit scoring. Entropy **22**(9), 989 (2020)
32. González, S., García, S., Del Ser, J., Rokach, L., Herrera, F.: A practical tutorial on bagging and boosting based ensembles for machine learning: algorithms, software tools, performance study, practical perspectives and opportunities. Inf. Fusion **64**, 205–237 (2020)
33. Gulsoy, N., Kulluk, S.: A data mining application in credit scoring processes of small and medium enterprises commercial corporate customers. Wiley Interdiscip. Rev. Data Min. Knowl. Discov. **9**(3), e1299 (2019)
34. Gunnarsson, B.R., Vanden Broucke, S., Baesens, B., Óskarsdóttir, M., Lemahieu, W.: Deep learning for credit scoring: do or don't? Eur. J. Oper. Res. **295**(1), 292–305 (2021)
35. Guo, S., He, H., Huang, X.: A multi-stage self-adaptive classifier ensemble model with application in credit scoring. IEEE Access **7**, 78549–78559 (2019)
36. Gupta, B., Dhawan, S.: Deep learning research: scientometric assessment of global publications output during 2004–17. Emerg. Sci. J. **3**(1), 23–32 (2019)
37. Hayashi, Y.: Emerging trends in deep learning for credit scoring: a review. Electronics **11**(19), 3181 (2022)
38. How, D.N.T., Loo, C.K., Sahari, K.S.M.: Behavior recognition for humanoid robots using long short-term memory. Int. J. Adv. Rob. Syst. **13**(6), 1729881416663369 (2016)
39. Hussin Adam Khatir, A.A., Bee, M.: Machine learning models and data-balancing techniques for credit scoring: what is the best combination? Risks **10**(9) (2022). https://doi.org/10.3390/risks10090169
40. Jadhav, S., He, H., Jenkins, K.: Information gain directed genetic algorithm wrapper feature selection for credit rating. Appl. Soft Comput. **69**, 541–553 (2018)
41. Jiao, W., Hao, X., Qin, C.: The image classification method with CNN-XGBoost model based on adaptive particle swarm optimization. Information **12**(4), 156 (2021)
42. Karalis, G.: Decision trees and applications. In: GeNeDis 2018: Computational Biology and Bioinformatics, pp. 239–242 (2020)

43. Kazemi, H.R., Khalili-Damghani, K., Sadi-Nezhad, S.: Tuning structural parameters of neural networks using genetic algorithm: a credit scoring application. Expert Syst. **38**(7), e12733 (2021). https://doi.org/10.1111/exsy.12733. https://onlinelibrary.wiley.com/doi/abs/10.1111/exsy.12733

44. Korangi, K., Mues, C., Bravo, C.: A transformer-based model for default prediction in mid-cap corporate markets. Eur. J. Oper. Res. **308**(1), 306–320 (2023)

45. Koutanaei, F.N., Sajedi, H., Khanbabaei, M.: A hybrid data mining model of feature selection algorithms and ensemble learning classifiers for credit scoring. J. Retail. Consum. Serv. **27**, 11–23 (2015)

46. Kumar, A., Sharma, S., Mahdavi, M.: Machine learning (ML) technologies for digital credit scoring in rural finance: a literature review. Risks **9**(11), 192 (2021)

47. Lee, T.S., Chiu, C.C., Lu, C.J., Chen, I.F.: Credit scoring using the hybrid neural discriminant technique. Expert Syst. Appl. **23**(3), 245–254 (2002)

48. Lei, K., Xie, Y., Zhong, S., Dai, J., Yang, M., Shen, Y.: Generative adversarial fusion network for class imbalance credit scoring. Neural Comput. Appl. **32**, 8451–8462 (2020)

49. Li, G., Ma, H.D., Liu, R.Y., Shen, M.D., Zhang, K.X.: A two-stage hybrid default discriminant model based on deep forest. Entropy **23**(5), 582 (2021)

50. Li, Y., Chen, W.: A comparative performance assessment of ensemble learning for credit scoring. Mathematics **8**(10), 1756 (2020)

51. Lin, C., Qiao, N., Zhang, W., Li, Y., Ma, S.: Default risk prediction and feature extraction using a penalized deep neural network. Stat. Comput. **32**(5), 76 (2022)

52. Liu, W., Fan, H., Xia, M.: Credit scoring based on tree-enhanced gradient boosting decision trees. Expert Syst. Appl. **189**, 116034 (2022)

53. Liu, W., Fan, H., Xia, M.: Step-wise multi-grained augmented gradient boosting decision trees for credit scoring. Eng. Appl. Artif. Intell. **97**, 104036 (2021)

54. Liu, W., Fan, H., Xia, M.: Multi-grained and multi-layered gradient boosting decision tree for credit scoring. Appl. Intell. **52**, 1–17 (2021)

55. Liu, Z., Pan, S.: Fuzzy-rough instance selection combined with effective classifiers in credit scoring. Neural Process. Lett. **47**, 193–202 (2018)

56. Livieris, I.E.: Forecasting economy-related data utilizing weight-constrained recurrent neural networks. Algorithms **12**(4), 85 (2019)

57. Louzada, F., Ara, A., Fernandes, G.B.: Classification methods applied to credit scoring: systematic review and overall comparison. Surv. Oper. Res. Manag. Sci. **21**(2), 117–134 (2016)

58. Luo, C., Wu, D., Wu, D.: A deep learning approach for credit scoring using credit default swaps. Eng. Appl. Artif. Intell. **65**, 465–470 (2017)

59. Markov, A., Seleznyova, Z., Lapshin, V.: Credit scoring methods: latest trends and points to consider. J. Financ. Data Sci. **8**, 180–201 (2022)

60. Marques, A., García, V., Sánchez, J.S.: A literature review on the application of evolutionary computing to credit scoring. J. Oper. Res. Soc. **64**, 1384–1399 (2013)

61. Mishra, C., Gupta, D.: Deep machine learning and neural networks: an overview. IAES Int. J. Artif. Intell. **6**(2), 66 (2017)

62. Moo-Young, M.: Comprehensive Biotechnology. Elsevier, Amsterdam (2019)

63. Munkhdalai, L., Munkhdalai, T., Ryu, K.H.: GEV-NN: a deep neural network architecture for class imbalance problem in binary classification. Knowl.-Based Syst. **194**, 105534 (2020)

64. Munkhdalai, L., Wang, L., Park, H.W., Ryu, K.H.: Advanced neural network app-roach, its explanation with LIME for credit scoring application. In: Nguyen, N.T., Gaol, F.L., Hong, T.-P., Trawiński, B. (eds.) ACIIDS 2019, Part II. LNCS (LNAI), vol. 11432, pp. 407–419. Springer, Cham (2019). https://doi.org/10.1007/978-3-030-14802-7_35
65. Nasreen, G., Haneef, K., Tamoor, M., Irshad, A.: A comparative study of state-of-the-art skin image segmentation techniques with CNN. Multimedia Tools Appl. **82**(7), 10921–10942 (2023)
66. Niu, B., Ren, J., Li, X.: Credit scoring using machine learning by combing social network information: evidence from peer-to-peer lending. Information **10**(12), 397 (2019)
67. Ozbayoglu, A.M., Gudelek, M.U., Sezer, O.B.: Deep learning for financial appli-cations: a survey. Appl. Soft Comput. **93**, 106384 (2020)
68. Pandey, M.K., Mittal, M., Subbiah, K.: Optimal balancing & efficient feature ranking approach to minimize credit risk. Int. J. Inf. Manag. Data Insights **1**(2), 100037 (2021)
69. Pavlyshenko, B.: Using stacking approaches for machine learning models. In: 2018 IEEE Second International Conference on Data Stream Mining & Processing (DSMP), pp. 255–258. IEEE (2018)
70. Pławiak, P., Abdar, M., Acharya, U.R.: Application of new deep genetic cascade ensemble of SVM classifiers to predict the Australian credit scoring. Appl. Soft Comput. **84**, 105740 (2019)
71. Provenzano, A.R., et al.: Machine learning approach for credit scoring. arXiv preprint arXiv:2008.01687 (2020)
72. Qian, H., Ma, P., Gao, S., Song, Y.: Soft reordering one-dimensional convolutional neural network for credit scoring. Knowl.-Based Syst. **266**, 110414 (2023)
73. Sadok, H., Sakka, F., El Maknouzi, M.E.H.: Artificial intelligence and bank credit analysis: a review. Cogent Econ. Financ. **10**(1), 2023262 (2022)
74. Sagi, O., Rokach, L.: Ensemble learning: a survey. Wiley Interdiscip. Rev. Data Min. Knowl. Discov. **8**(4), e1249 (2018)
75. Schapire, R.E.: The boosting approach to machine learning: an overview. In: Deni-son, D.D., Hansen, M.H., Holmes, C.C., Mallick, B., Yu, B. (eds.) Nonlinear Estimation and Classification, vol. 171, pp. 149–171. Springer, New York (2003). https://doi.org/10.1007/978-0-387-21579-2_9
76. Shahraki, A., Abbasi, M., Haugen, Ø.: Boosting algorithms for network intru-sion detection: a comparative evaluation of real AdaBoost, gentle AdaBoost and modest AdaBoost. Eng. Appl. Artif. Intell. **94**, 103770 (2020)
77. Shen, F., Zhao, X., Kou, G., Alsaadi, F.E.: A new deep learning ensemble credit risk evaluation model with an improved synthetic minority oversampling tech-nique. Appl. Soft Comput. **98**, 106852 (2021)
78. Shen, F., Zhao, X., Lan, D., Ou, L.: A hybrid model of AdaBoost and back-propagation neural network for credit scoring. In: Xu, J., Gen, M., Hajiyev, A., Cooke, F.L. (eds.) ICMSEM 2017. LNMIE, pp. 78–90. Springer, Cham (2018). https://doi.org/10.1007/978-3-319-59280-0_6
79. Shi, S., Tse, R., Luo, W., D'Addona, S., Pau, G.: Machine learning-driven credit risk: a systemic review. Neural Comput. Appl. **34**(17), 14327–14339 (2022)
80. Tang, L., Cai, F., Ouyang, Y.: Applying a nonparametric random forest algorithm to assess the credit risk of the energy industry in China. Technol. Forecast. Soc. Chang. **144**, 563–572 (2019)

81. Teles, G., Rodrigues, J.J., Saleem, K., Kozlov, S., Rabêlo, R.A.: Machine learning and decision support system on credit scoring. Neural Comput. Appl. **32**, 9809–9826 (2020)

82. Teply, P., Polena, M.: Best classification algorithms in peer-to-peer lending. N. Am. J. Econ. Finance **51**, 100904 (2020)

83. Tiwari, A.: Introduction to machine learning. Ubiquitous Mach. Learn. Its Appl. (2017)

84. Tripathi, D., Edla, D.R., Kuppili, V., Bablani, A.: Evolutionary extreme learning machine with novel activation function for credit scoring. Eng. Appl. Artif. Intell. **96**, 103980 (2020)

85. Tripathi, D., Edla, D.R., Kuppili, V., Dharavath, R.: Binary BAT algorithm and RBFN based hybrid credit scoring model. Multimedia Tools Appl. **79**, 31889–31912 (2020)

86. Tripathi, D., Shukla, A.K., Reddy, B.R., Bopche, G.S., Chandramohan, D.: Credit scoring models using ensemble learning and classification approaches: a comprehensive survey. Wireless Pers. Commun. **123**, 1–28 (2022)

87. Veeramanikandan, V., Jeyakarthic, M.: Parameter-tuned deep learning model for credit risk assessment and scoring applications. Recent Adv. Comput. Sci. Commun. (Formerly: Recent Patents on Computer Science) **14**(9), 2958–2968 (2021)

88. Wang, C., Han, D., Liu, Q., Luo, S.: A deep learning approach for credit scoring of peer-to-peer lending using attention mechanism LSTM. IEEE Access **7**, 2161–2168 (2018)

89. Wang, G., Hao, J., Ma, J., Jiang, H.: A comparative assessment of ensemble learning for credit scoring. Expert Syst. Appl. **38**(1), 223–230 (2011)

90. Wang, Y., Jia, Y., Zhong, Y., Huang, J., Xiao, J.: Balanced incremental deep reinforcement learning based on variational autoencoder data augmentation for customer credit scoring. Eng. Appl. Artif. Intell. **122**, 106056 (2023)

91. Wei, S., Yang, D., Zhang, W., Zhang, S.: A novel noise-adapted two-layer ensemble model for credit scoring based on backflow learning. IEEE Access **7**, 99217–99230 (2019)

92. Witten, I.H., Frank, E., Hall, M.A., Pal, C.J., Data, M.: Practical machine learning tools and techniques. In: Data Mining, vol. 2 (2005)

93. Wong, M.L., Seng, K., Wong, P.K.: Cost-sensitive ensemble of stacked denoising autoencoders for class imbalance problems in business domain. Expert Syst. Appl. **141**, 112918 (2020)

94. Wu, C.F., Huang, S.C., Chiou, C.C., Wang, Y.M.: A predictive intelligence system of credit scoring based on deep multiple kernel learning. Appl. Soft Comput. **111**, 107668 (2021)

95. Xia, Y., Liu, C., Da, B., Xie, F.: A novel heterogeneous ensemble credit scoring model based on bstacking approach. Expert Syst. Appl. **93**, 182–199 (2018)

96. Xiao, J., et al.: A novel deep ensemble model for imbalanced credit scoring in internet finance. Int. J. Forecast. **40**, 348–372 (2023)

97. Xu, D., Zhang, X., Feng, H.: Generalized fuzzy soft sets theory-based novel hybrid ensemble credit scoring model. Int. J. Finance Econ. **24**(2), 903–921 (2019)

98. Xu, D., Zhang, X., Hu, J., Chen, J.: A novel ensemble credit scoring model based on extreme learning machine and generalized fuzzy soft sets. Math. Probl. Eng. **2020** (2020)

99. Yobas, M.B., Crook, J.N., Ross, P.: Credit scoring using neural and evolutionary techniques. IMA J. Manag. Math. **11**(2), 111–125 (2000)

100. Yotsawat, W., Wattuya, P., Srivihok, A.: A novel method for credit scoring based on cost-sensitive neural network ensemble. IEEE Access **9**, 78521–78537 (2021)

101. Yu, L., Zhou, R., Tang, L., Chen, R.: A DBN-based resampling SVM ensemble learning paradigm for credit classification with imbalanced data. Appl. Soft Comput. **69**, 192–202 (2018)
102. Zanin, L.: Combining multiple probability predictions in the presence of class imbalance to discriminate between potential bad and good borrowers in the peer-to-peer lending market. J. Behav. Exp. Financ. **25**, 100272 (2020)
103. Zhang, H., He, H., Zhang, W.: Classifier selection and clustering with fuzzy assignment in ensemble model for credit scoring. Neurocomputing **316**, 210–221 (2018)
104. Zhang, W., Xu, W., Hao, H., Zhu, D.: Cost-sensitive multiple-instance learning method with dynamic transactional data for personal credit scoring. Expert Syst. Appl. **157**, 113489 (2020)
105. Zhang, Z., Niu, K., Liu, Y.: A deep learning based online credit scoring model for P2P lending. IEEE Access **8**, 177307–177317 (2020)
106. Zhong, Y., Wang, H.: Internet financial credit scoring models based on deep forest and resampling methods. IEEE Access **11**, 8689–8700 (2023)
107. Zhu, F., Chen, X., Li, G.: Multi-classification assessment of personal credit risk based on stacking integration. Procedia Comput. Sci. **214**, 605–612 (2022)

Refining Detection Mechanism of Mobile Money Fraud Using MoMTSim Platform

Denish Azamuke$^{(\boxtimes)}$ [ID], Marriette Katarahweire [ID], Joshua Muleesi Businge [ID], Samuel Kizza [ID], Chrisostom Opio [ID], and Engineer Bainomugisha [ID]

Department of Computer Science, Makerere University, Plot 56 Pool Road, Kampala, Uganda
{azamuke.denish,joshuamuleesi.businge,samuel.kizza25, chrisostom.opio}@students.mak.ac.ug, {marriette.katarahweire,baino}@mak.ac.ug
https://cs.mak.ac.ug/

Abstract. Mobile money financial crime evolves in various forms including account takeover fraud, refund fraud, and fake credentials, which poses challenges in measuring its overall cost using real transaction data. Current machine learning methods struggle due to outdated historical financial data that cannot be used to study emerging fraud patterns. This paper introduces MoMTSim, a virtual mobile money platform developed and calibrated using real mobile money transaction data from Sub-Saharan Africa to refine fraud detection mechanisms. Employing statistical methods, agent-based modeling, and social network analysis, MoMTSim improves fidelity by comparing real and synthetic data using the sum of squared errors (SSE) approach. Our experiments show consistent success in fraud classification using machine learning algorithms inclusive of Random Forest and XGBoost. Simpler models encompassing Logistic Regression, KNN, and Decision Trees also exhibit remarkable performance in mobile money fraud classification. This approach aids in studying new fraud scenarios and strengthening financial fraud detection mechanisms using the virtual mobile money platform.

Keywords: Mobile money · Agent-based modeling · Machine learning · Financial fraud

1 Introduction

Mobile money financial fraud is a challenge that manifests in various forms [2,4, 24]. Mobile money services are widely used in Sub-Saharan Africa. The impact of fraudulent activity on the mobile money platform can be substantial and calls for massive investment in fraud detection and prevention [1,2]. This investment in mobile money security could effectively be implemented if the true cost of fraud can be measured [27,28]. The difficulty in measuring the total cost of financial fraud using real transaction datasets is due to ever-emerging fraud patterns.

T. G. Debelee et al. (Eds.): PanAfriConAI 2023, CCIS 2069, pp. 62–82, 2024.
https://doi.org/10.1007/978-3-031-57639-3_3

New fraudulent scenarios, including account takeover fraud, refund fraud, and fake credentials are common in the real ecosystem [2]. Account takeover fraud is concerned with a fraudster gaining unauthorized access to a victim's account in the form of SIM swaps and stolen credentials. The use of machine learning techniques for fraud detection is ineffective due to the deprecated nature of the historical data for studying patterns of unfolding fraud schemes [27].

A recent study [24] compares the performance metrics of different classification algorithms on mobile money data including Logistic Regression, Random Forest, and Decision Trees. The study relied on a synthetic dataset containing a single fraud pattern simulated using the PaySim [26] financial simulator. PaySim is capable of generating synthetic financial data however, the data model lacks crucial client attributes consisting of email, national identification number (NIN), and phone number among other things. These client attributes form the basis for modeling common fraud scenarios arising from fake credentials. Moreover, previous studies [2,3,22,24,25] did not provide a comparison of cost implications for the common fraud scenarios. Comparing different machine learning (ML) algorithms based on model performance evaluation metrics and estimates for the total cost of fraud makes it possible to understand financial losses due to hidden fraud [27,28].

Our approach to financial fraud modeling and detection leverages security threat attributes to combat mobile money fraud challenges, utilizing simulation and computational methods for behavioral detection. This study uses the MoMTSim financial simulation platform to measure the total cost of losses through the simulation of fraudulent scenarios, including account takeover fraud, to enrich the synthetic data. MoMTSim [13] is a multi-agent-based simulation (MABS) platform, a virtual mobile money service based on real mobile money transaction data and properties of the real ecosystem found in Sub-Saharan Africa. Scenario-based and social network analysis of customer-to-customer and customer-to-merchant interactions were crucial during the design of the simulation model. Real mobile money transaction dataset was used to develop and calibrate the simulation model of MoMTSim. The fidelity of the generated data was measured using statistical methods encompassing the sum of squared errors (SSE) approach. The SSE method computes the delta between the real and synthetic data. MoMTSim yields diverse synthetic mobile money transaction data due to the low-level interactions of the agents (agents mirror real-world actors).

We performed experiments, including an assessment of the statistical similarity of the synthetic data to the real data. We computed estimates for the total cost of fraud in mobile money transactions and performed financial fraud classification, focusing on comparing the performance of different machine learning algorithms for fraud detection. The experiments provide insights into automated fraud detection using simulation and computational approaches. In comparison, deterministic rule-based methods often fail to identify suspicious transactions and complex fraud patterns. Moreover, fraud simulation and synthetic datasets allow the study of anticipated fraudulent activities and patterns that do not exist in the real data. This ultimately offers limitless ways for law enforcement authorities and service providers to fine-tune their fraud control methods and systems [27,28].

This paper makes the following contributions.

– We determine the statistical similarity of the synthetic transaction data generated using MoMTSim to the real data.
– The study evaluates existing fraud control methods composed of a lower and an upper cap on transfer transactions and estimates the total cost of fraud in mobile money transactions including the hidden fraud that is non-obvious to measure.
– The study performs financial fraud classification and compares the performance of the different ML algorithms, thereby suggesting the Random Forest and XGBoost algorithms for detecting fraud in mobile money transactions.

The rest of the paper is organized as follows: Sect. 2 presents fraudulent activities in mobile money services. The method and experiments carried out are discussed in Sect. 3. Section 4 discusses the results of the experiments. Related work in financial fraud detection using machine learning algorithms is presented in Sect. 5, and the paper makes conclusions in Sect. 6.

2   Fraudulent Activities in Mobile Money Services

Mobile money systems in Sub-Saharan Africa face varying fraud challenges, with some being more prevalent. This study focuses on specific financial fraud challenges inclusive of account takeover fraud, refund fraud, and fake credentials that were identified through experts' opinions. The opinions of domain experts also guided our experimental approach in this study while acknowledging the existence of other fraud scenarios in the real mobile money ecosystem. This study concentrated on fraud scenarios deemed significant by domain experts for modeling to build computational methods for the increasing fraud challenges in mobile money transactions.

2.1   Account Takeover Fraud

A notorious fraud scenario in the mobile money ecosystem is account takeover fraud in the form of SIM swaps and lost credentials [2,4]. A SIM swap fraud involves a fraudster replacing a customer's SIM card, thereby gaining control of the victim's phone number associated with their mobile money account. This allows the fraudster to execute unauthorized transactions using the victim's account. The fraudster is capable of emptying the victim's account by performing numerous transactions that are eventually withdrawn from the mobile money system through a mobile money merchant. Lost credentials occur when a customer loses their phone, enabling fraudsters to steal their mobile money PIN. Mobile money account access is protected by a short PIN code, which the fraudster can change once they have gained control of the victim's device [2,4].

2.2   Refund Fraud

A refund fraud scenario entails the fraudster making a payment for goods or services and subsequently requesting multiple refunds. The fraudster identifies susceptible merchants, conducts numerous transactions, including those with potential new victims, and ultimately executes withdrawals either through their account or with the assistance of mules [4].

2.3   Fake Credentials

A fake credential fraud scenario involves creating a mobile money account using another person's phone number, email, or national identification number (NIN). The fraudster utilizes this account to conduct fraudulent transactions on behalf of the individual associated with the used identification information [4].

3   Method and Experiments

3.1   Synthetic Data Generation Using MoMTSim

Before the simulation of fraudulent behaviors to enhance synthetic data, this study introduces the MoMTSim financial simulation platform, employing a generic MABS toolkit, MASON [29]. MASON is fast enough, core in Java, and multi-platform. It supports parallelism and is capable of reinforcing computationally intensive simulations [30] unlike other simulation toolkits encompassing NetLogo [44], RePast [12], and AnyLogic [19]. MoMTSim has many agents based on the characteristics of actors (banks, clients, fraudsters, mobile money merchants) in the real world [27,28]. The simulation model in MoMTSim has the client as the main entity with an associated client profile, the number of transactions the client can make within a given period, limits on the transactions, and the initial balance distributions. Clients carry out transactions, including deposits, transfers, payments, debits, and withdrawals. A record of every transaction made by a client is kept within the system (the virtual world). Client participation in a transaction is determined using a random variable contingent on calculated probabilities from the analysis of the real data [26]. Therefore, the state of a client is alterable in the simulations. Other entities in the simulation model include a mobile money merchant who supports clients in deposit and withdrawal transactions and a bank that accepts mobile money transactions. Some clients are designed to act fraudulently, mimicking the behaviors of fraudsters in the real ecosystem.

MoMTSim includes several parameters to exert precise control over the simulations. The parameters encompass the total number of each agent category, the mapping of time, and probabilities for reusing agents in future transactions. Input file paths of CSV files for aggregated mobile money transactions, client profiles, initial account balance distributions, transaction types, and limits on mobile money transfers, among other things, are specified. Additionally, an output file path to the storage space for the synthetic datasets and other log files after a complete simulation is defined. In the simulation, one step represents one hour in the real world. These parameters are crucial for fine-tuning and customizing the simulation process to ensure its accurate representation of real-world scenarios. They also enable the scaling of simulations to an arbitrarily large number of agents and transactions. Therefore, a successful simulation in MoMTSim undergoes the following processes.

- First, input files for a simulation are loaded, comprising parameter files containing values for each parameter. These parameters include the file paths for source data inputs, statistical distributions for each transaction type, and the starting balance for every client. A starting balance is then assigned for every client generated during the simulation and mobile money merchants are initialized, signifying that they are set up and prepared within the simulation environment [27,28]. This ensures that clients and mobile money merchants are ready to participate in the simulation.
- After input files are loaded, MoMTSim proceeds to the actual simulation to produce synthetic data. MoMTSim converts a step in a simulation into a day/hour pairing, used as input to derive statistical distributions from the aggregated mobile money transaction file. A probability q for executing every transaction in the simulation is extracted and incorporated into the client's model resulting in a client becoming aware of the transaction count, participation in future steps of the simulation, probability distribution for carrying a given transaction, and their starting account balance [26,28]. Upon creation of a client, they carry out transactions based on the loaded distributions with other clients, while mobile money merchants assume a passive role of aiding deposit and withdrawal transactions for clients.
- Following the completion of agent interactions in MoMTSim (for every complete simulation), the log file, CSV file of raw mobile money transactions, aggregated output resembling the real aggregated transaction data, and parameter file history are saved on the device.

The simulation model was calibrated to ensure that every agent in MoMTSim is believable based on documented properties of mobile money transactions [2,4], expert opinions, transaction patterns in the real data, and our understanding of the real mobile money financial ecosystem in the region. Parameter sets leading to unusual behavior of agents were removed and the fidelity of the synthetic data was evaluated using the sum of squared errors (SSE) method [27,28] by computing the difference between the real and synthetic data. The dataset with the least total error was chosen for further analysis and the aggregated transaction data for real and synthetic datasets were used to compare the statistical properties of the data [26–28].

Our simulation-based approach to detecting mobile money transaction fraud using the MoMTSim platform is shown in Fig. 1. The simulation process is iterative, involving the calibration of several parameters of the agent model. Once the simulation model is deemed capable of generating synthetic datasets that statistically conform to real data, fraudulent behaviors can be introduced in the simulation platform. The injection of fraudulent behaviors involves introducing new parameters in the simulation model per fraud scenario discussed in Sect. 2 to augment the synthetic mobile money transaction datasets for fraud detection. We discuss experiments aimed at estimating the cost of mobile money transaction fraud and financial fraud classification using various machine learning algorithms in detail in Subsects. 3.2 and 3.3, respectively.

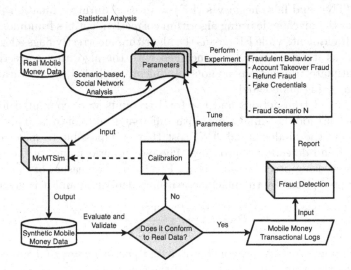

Fig. 1. An iterative simulation-based approach to fraud detection in mobile money transactions using the MoMTSim platform.

3.2 Estimation of the Cost of Fraud in Mobile Money Transactions

Estimating the fraud cost in mobile money transactions involved enriching the synthetic data using the account takeover fraud scenario and employing the resulting datasets to assess the existing mobile money fraud controls. We introduced a fraud parameter targeting account takeover fraud (discussed in Subsect. 2.1), defining a probability that ensures the effective generation of sufficient fraudulent transactions. We then formulated transaction rules designed to mimic fraudulent activity, characterized by a series of rapid transactions aimed at depleting the balance of the compromised account [4]. This illicit activity also involves using intermediary accounts or mules to facilitate the withdrawal of the pilfered funds from the mobile money platform. On completion of the injection of fraudulent scenario in MoMTSim, we introduced caps (limits) on transfer transactions. These limits are existing fraud controls on transfer transactions in the real ecosystem (also known as transfer limits) as guided by domain experts from regional financial institutions. Besides, Lopez-Rojas et al. [28] explored intermediary values between the lower and upper caps. However, our study did not replicate this approach but instead concentrated on new caps determined through consultations with domain experts. The first control was a lower transfer transaction limit of 100,000 virtual units[1]. And, we simulated sufficient instances of account takeover fraud with it. The simulation process was repeated with a higher transfer limit of 10,000,000 virtual units. The resulting datasets were used to train the Logistic Regression model (base model) for fraud detection, and the key metrics encompassing true positives (TP), false positives (FP), true negatives (TN), and false negatives (FN) were used during evaluation [17]. TP implies that the machine learning algorithm correctly identifies fraudulent transactions as fraudulent, while FP means the algorithm incorrectly flags a legitimate transaction as fraudulent. TN is concerned with the algorithm correctly identifying legitimate transactions as non-fraudulent, while FN involves fraudulent transactions going undetected.

The count of transactions and the total amounts were computed for all of the key metrics for the simulations with different fraud controls. This allowed us to measure the hidden fraud (FN) cost that is impossible to measure using traditional fraud detection approaches. The model performance of the classifier was evaluated using precision and recall. Precision is a measure of the accuracy of positive predictions by the machine learning algorithm, and it is given by

$$\text{Precision} = \frac{TP}{TP + FP}. \tag{1}$$

While recall measures the proportion of actual positive cases that are correctly identified by the algorithm. It is given by the equation

$$\text{Recall} = \frac{TP}{TP + FN}. \tag{2}$$

The results for the two controls used were compared based on the model performance evaluation metrics and the estimates for the cost of fraud.

3.3 Financial Fraud Detection Using Machine Learning Algorithms

A new dataset with refund fraud and transactions resulting from accounts with fake credentials (discussed in Subsects. 2.2 and 2.3) was simulated for financial fraud classification using various machine learning algorithms. Similar to account takeover fraud, specific parameters, and transaction rules were defined for each fraud scenario.

Model Selection and Training. Machine learning models considered for fraud classification included Logistic Regression [21], Random Forest [34], XGBoost [9], Decision Trees [7] and KNN [15]. These algorithms have proven capabilities

---

[1] In this paper, "Virtual units" (abbreviated as "VUs") are used to represent financial transactions in a Sub-Saharan mobile money platform. This is due to a non-disclosure agreement that restricts disclosure of the actual name of the currency.

in financial crime detection, ranging from the well-established dependability of Logistic Regression and the adaptability of Random Forest to the easily understandable nature of Decision Trees among other things [6,37,38,43]. They provide comprehensive and versatile mechanisms for detecting complex financial crimes [6,37,38,43]. The Logistic Regression model was used as a baseline model for the task due to its simplicity and ease of training.

A labeled dataset with independent features and the dependent variable (isFraud), shown in Table 1, was generated. Client identifiers were omitted before model building since they would not affect the results of the experiment. The features in the simulated data had varying scales. To smooth the model training process, the min-max scaler method was used for feature scaling since mobile money transactions do not follow the normal distribution. It is given by the equation

$$\text{Scaled value} = \frac{X - X_{\min}}{X_{\max} - X_{\min}}, \tag{3}$$

for feature X in the dataset.

The simulated data contained no missing values and it had sufficient instances of fraudulent transactions. SMOTE [8] was therefore not used to address the class imbalances that are often common with financial data. This is unarguably one of the advantages of using synthetic data over real data for financial fraud detection [4].

The synthetic dataset contained 3,986,294 rows of transactions of which 79.4% were legitimate transactions while 20.6% were fraudulent transactions. It was divided into training and test sets to assess the model performance while also conducting model training and testing experiments on distinct data samples [20,31]. The training set had 70% of the data while the test set contained 30% with both classes (legitimate and fraudulent transactions) included in each set.

Model Performance Evaluation. Precision and recall are given by Eqs. (1) and (2) respectively. In the fraud classification experiment, we also used other model performance evaluation metrics consisting of the Matthews Correlation Coefficient (MCC), F1-score, and the AUC-ROC. MCC measures the correlation of the true classes c with the predicted labels l with +1 representing a perfect prediction, 0 being an average random prediction while -1 indicates a completely inverse prediction. It serves as a balanced metric for classes with different sizes [10,11,36] including financial data. It is given by

$$\text{MCC} = \frac{Cov(c,l)}{\sigma_c \sigma_l} = \frac{(\text{TP} \times \text{TN} - \text{FP} \times \text{FN})}{\sqrt{(\text{TP} + \text{FP}) \times (\text{TP} + \text{FN}) \times (\text{TN} + \text{FP}) \times (\text{TN} + \text{FN})}}, \tag{4}$$

where $Cov(c,l)$ is the covariance of the true classes c and predicted labels l, σ_c and σ_l are the standard deviations, respectively. F1-score is the harmonic mean of precision and recall, a more informative model performance evaluation metric

Table 1. The independent features and the dependent variable in the synthetic mobile money transaction data.

Feature/Variable	Description	Measure
step	This maps a unit of time, a step in the simulation is equivalent to one hour in the real world	Continuous
transactionType	Transaction types in the simulations include deposit, withdrawal, transfer, debit, and payment. They are the operations that also result in changes in the account balance of a client, and the movement of funds from one client account to another account	Categorical
amount	This is e-money associated with a transaction type	Continuous
startingClient	Mobile money customer who initiates a transaction	Continuous
oldBalStartingClient	The starting balance of a client initiating a transaction before the transaction happens	Continuous
newBalStartingClient	The new balance of a client initiating a transaction after performing the transaction	Continuous
destinationClient	The recipient of funds after a transaction has taken place	Continuous
oldBalDestinationClient	The initial balance of the recipient client before a transaction is delivered to their account	Continuous
newBalDestinationClient	The new balance of a recipient client after a transaction has happened	Continuous
isFraud	Label for a transaction, 1 for a fraudulent transaction, and 0 for a legitimate transaction	Categorical

than accuracy, given by

$$\text{F1-score} = 2 \times \frac{\text{Precision} \times \text{Recall}}{\text{Precision} + \text{Recall}}. \tag{5}$$

The optimal F1-score is 1 implying a perfect precision and recall while 0 otherwise. With the number of fraudulent transactions significantly lower than that of legitimate transactions typically in financial datasets, the F1-score is preferred to the accuracy metric [24] when it comes to identifying fraudulent transactions while minimizing the number of false alarms. However, we also computed the overall accuracy scores for the models, given by the equation

$$\text{Accuracy} = \frac{\text{TP} + \text{TN}}{\text{TP} + \text{TN} + \text{FP} + \text{FN}}. \tag{6}$$

The ROC curve is a plot for the true positive rate (TPR) against the false positive rate (FPR) at various threshold values. The ROC curve closer to the left corner of the graph indicates a good classification model while the one closer to the diagonal line represents a random model. The area under the ROC curve (AUC-ROC) was also used to compare the different classification algorithms [6] with values ranging from 0 to 1 whereby a value closer to 1 shows good prediction while a value closer to 0 shows poor prediction.

4   Results and Discussion

A quartet of synthetic datasets were generated for the analysis of the results.

4.1   Statistical Similarity of the Datasests

First, to show that the synthetic datasets generated resemble the real data, we used the aggregated transactions of MoMTSim_202308 simulation and one for the real dataset to carry out statistical comparisons. MoMTSim_202308 implies synthetic data generated using the MoMTSim financial simulation platform in the year 2023 and for August.

The density distributions of the real and synthetic data were plotted, with each distribution representing the total transaction values for the transaction types. These distributions were overlapped to visually compare how the total transaction values for each transaction type behaved in both the real and synthetic datasets. Figure 2 shows overlapping density distributions of the datasets with some differences in the tails and peak densities. The density plots for deposit, withdrawal, payment, and transfer transactions exhibit significant overlap with shapes in the synthetic data closely resembling those in the real data. Similarly, the shapes of the debit transactions in the real and synthetic data closely resemble each other. The raw simulated transaction data of MoMTSim_202308 contained about 34 million transaction records with 42.75% deposits, followed by 24.04% withdrawal transactions, 23.66% payments, 5.99% transfers,

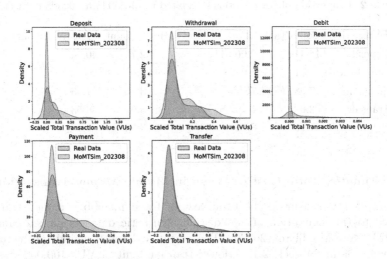

Fig. 2. Density comparison of scaled total transaction values by transaction type between real and synthetic data.

and 3.56% debit transactions. In terms of the total transaction value per transaction type, deposits had the largest total amount taking 40.94%, followed by transfers with 36.53%, 20.92% withdrawals, 1.43% payments, and 0.18% debit transactions. Deposits are concerned with loading electronic money onto the mobile money platform. They form the basis for other transactions to happen within the platform. Transfers include intermediary transactions within the mobile money system, involving users sending money among themselves. Typically in the Sub-Saharan region, mobile money is often used as a transfer service which explains the high total amount for the transfer transactions. On the other hand, few users often send money from their mobile money account to a bank account due to the charges incurred in doing such transactions and this results in the low total amount for debits. Similarly, the adoption of mobile money payments has not gained traction in the region hence low total transaction value. Withdrawals simply allow a user to get money off their account as hard cash with the help of a mobile money merchant who facilitates the conversion of the mobile money funds. The percentage count of the simulated transactions and their corresponding total amounts are shown in Table 2. The unit M in Table 2 implies millions of transactions while B implies billions of VUs. Our evaluation of the statistical resemblance between the datasets aligns with findings from pertinent studies [3,27,28]. Hence, the similarity in density distributions between real and synthetic data demonstrates that MoMTSim generates datasets with statistical properties closely resembling real mobile money transactions, rendering them suitable for research purposes. During the simulations, we noticed some causal dynamics including an increase in activity during some steps and a decrease in other times. This is consistent with the operations of the mobile money platform in the real world.

Table 2. Proportion of transactions simulated in MoMTSim_202308 dataset.

Transaction	Count (M)	Total Amount (B)	% Count	% Total Amount
Deposit	14.41	2,009.11	42.75	40.94
Withdrawal	8.11	1,026.67	24.04	20.92
Payment	7.98	70.13	23.66	1.43
Transfer	2.02	1,792.55	5.99	36.53
Debit	1.20	8.81	3.56	0.18

4.2   Estimated Total Cost of Fraud in Mobile Money Transactions

Experiments for measuring the total cost of fraud in mobile money transactions led to the simulation of two distinct synthetic datasets named MoMT-Sim_202307 and MoMTSim_202309. The synthetic datasets contained the fraudulent behavior in Subsect. 2.1. A lower transfer limit of VUs 100,000 was set for the MoMTSim_202307 simulation while a higher limit of VUs 10,000,000 for MoMTSim_202309. Even though equal numbers of agents were permitted for both simulations, we observed varying proportions of fraudulent transactions in the simulated data. MoMTSim_202307 with a strict limit on transfers registered 1.3% lower fraudulent transactions than the MoMTSim_202309 simulation. Sufficient amounts of fraudulent transactions were simulated in both cases with 25.9% and 27.2% fraud for MoMTSim_202307 and MoMTSim_202309 respectively.

After classifying fraud with the base model (Logistic Regression) using the dataset, MoMTSim_202307, with a strict limit of VUs 100,000 registered a higher number of false positives, specifically 321,465 transactions, totaling VUs 7.63 billion, in contrast to MoMTSim_202309. MoMTSim_202309 had a higher transfer limit of VUs 10,000,000 and registered 9,631 false positive transactions with a total value of VUs 0.28 billion. On the other hand, MoMTSim_202309 registered a higher number of false negatives (57,888 transactions) with a total value of VUs 1.43 billion as compared to MoMTSim_202307 with 10,982 transactions, totaling VUs 0.36 billion. These observations indicate that a strict transfer limit on mobile money transactions is likely to result in many false positives, hence inconveniencing clients conducting legitimate transactions. A flexible (higher) transfer limit is likely to result in more hidden fraud (fraudulent transactions that make it through the mobile money system).

With strict control (lower transfer limit), 195,333 fraudulent transactions with a total amount of 4.70 billion were correctly identified compared to the 62,738 true positive transactions with a total value of VUs 1.62 billion for the flexible control. The classifier correctly identified 2,108,979 legitimate transactions with a total transaction value of VUs 5.96 billion for the strict control while 2,212,470 normal transactions with VUs 8.44 billion total amount were correctly identified for the flexible control. This implies that in both cases, a relatively high number of true negative (normal) transactions were correctly detected by the algorithm. The estimates for the total cost of fraud in mobile money transactions using the two controls are shown in Table 3.

Table 4 shows the model performance metrics for the classifier. Strict control in MoMTSim_202307 led to a lower precision of 38% while a higher recall of 95%. With this control, a high number of false positives were registered while minimal false negatives were reported. The control is capable of minimizing hidden fraud (false negatives) at the cost of flagging legitimate transactions as fraudulent. With the flexible control (high transfer limit), the classifier registered a higher precision of 85% implying it minimizes flagging normal transactions as fraudulent but the recall (52%) is affected. With this fraud control, more fraudulent transactions pass through the mobile money system unnoticed and this affects the revenue even though clients are not inconvenienced with warnings of fraud. Therefore, the security teams at financial institutions or the service providers need to weigh the trade-offs between the two common fraud controls. They can probably settle for a balance where many legitimate customers are not inconvenienced while also not letting many fraudulent transactions go unnoticed through the mobile money system. These findings align with and corroborate the results observed in similar studies conducted in this research domain [27,28].

Table 3. Estimates for the cost of fraud using common fraud controls.

Simulation	MoMTSim_202307 (100K limit)		MoMTSim_202309 (10M limit)	
Class	Count	Total Amount (B)	Count	Total Amount (B)
TP	195,333	4.70	62,738	1.62
FP	321,462	7.63	9,631	0.28
TN	2,108,979	5.96	2,212,470	8.44
FN	10,982	0.36	57,888	1.43

Table 4. Model performance metrics with lower and upper caps on transfer transactions.

Simulation	Fraud Control	Precision	Recall
MoMTSim_202307	VUs 100,000 transfer limit	0.38	0.95
MoMTSim_202309	VUs 10,000,000 transfer limit	0.85	0.52

4.3 Financial Fraud Classification Results

MoMTSim_202310 dataset was simulated with two fraud scenarios (see Subsect. 2.2 and 2.3) to identify machine learning algorithms suitable for fraud detection in mobile money transactions based on model performance and the total cost of financial fraud. Among the machine learning classifiers used in this experiment, the Random Forest model registered an F1-score of 0.80 and an MCC of 0.77 as shown in the results for the testing set in Table 5. Rows in Table 5 show model performance scores and estimated total cost of fraud in mobile money transaction fraud classification. Key scores are underlined to highlight their significance, following established standards in financial crime analysis. 'LR' denotes Logistic Regression, used as a baseline model for comparing the performance of other models.

Even though the simpler models exhibited good model performance including Logistic Regression with an MCC of 0.75 and an F1-score of 0.79, their metric scores for the cost of fraud especially for the number of false positives and false negatives are undesirable when compared with Random Forest. Only 19 transactions with an insignificant total transaction amount were reported as false positives by the Random Forest model while 531 false positives were recorded with a total amount of VUs 0.02 billion by XGBoost that showed similar model performance. The Random Forest model minimized the hidden fraud (false negatives) to a total transaction amount of VUs 2.69 billion. Even though KNN reported a lower total transaction amount of VUs 2.24 billion for the hidden fraud compared to the Random Forest classifier, it identified 41,368 legitimate transactions as fraudulent, a higher number than the 19 false positives for the Random Forest model. The Random Forest algorithm excels at learning from datasets with a high number of features [23,24,32]. By amalgamating the predictions of numerous individual decision trees, the Random Forest algorithm mitigates overfitting and enhances generalization, demonstrating the power of ensemble learning [24,32]. XGBoost model with a similar performance to Random Forest supports parallel data processing making it quickly process large financial datasets. The XGBoost model is also flexible, permitting the definition of custom optimization objectives and evaluation criteria to fine-tune the model for specific needs [33,41,45]. The Logistic Regression, Decision Trees, Random Forest, and XGboost showed a high AUC of 0.95 indicating superior performance as compared to KNN with an AUC of 0.93. But still, the KNN model demonstrated effectiveness in separating fraudulent transactions from legitimate ones and these scores are shown in the AUC-ROC plot in Fig. 3. Among the array of models employed in the experiment, the Random Forest and XGBoost models distinguished themselves as preferred choices for effectively reducing false alarms (false positives) and uncovering concealed instances of fraud (false negatives) in the context of mobile money transactions. They excelled in the challenging task of identifying substantial numbers of fraudulent transactions.

Table 5. Financial fraud classification results for the different classifiers.

Metric	LR	Decision Tree	Random Forest	KNN	XGBoost
Estimates for the Total Cost of Fraud					
TP Count	222,229	217,886	215,645	234,452	217,274
FP Count	15,039	2,118	19	41,368	531
TN Count	857,248	870,169	872,268	830,919	871,756
FN Count	101,373	105,716	107,957	88,150	106,328
TP Amount (B)	7.22	7.92	7.88	8.38	7.92
FP Amount (B)	1.22	0.07	0.00	1.08	0.02
TN Amount (B)	39.51	40.65	40.71	39.65	40.71
FN Amount (B)	3.40	2.69	2.69	2.24	2.69
Model Performance Metrics					
Precision	0.94	0.99	1.00	0.85	1.00
Recall	0.69	0.67	0.67	0.73	0.67
F1-score	0.79	0.80	0.80	0.78	0.80
MCC	0.75	0.77	0.77	0.72	0.77
Accuracy	0.90	0.91	0.91	0.89	0.91

Nonetheless, it is worth noting that more straightforward models, including the Logistic Regression algorithm, also exhibited proficiency in mobile money fraud detection. These simpler models can serve as valuable alternatives, particularly in resource-constrained settings where computational complexity and resource demands might limit the use of more sophisticated algorithms while still offering significant utility in fraud prevention.

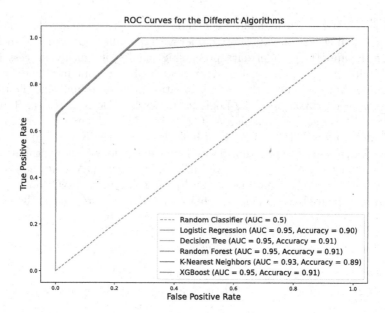

Fig. 3. Results from AUC-ROC.

5   Related Work

The work of Lokanan [24] compares different machine learning classifiers to predict financial fraud in mobile money transfers using a synthetic dataset generated with PaySim [26] financial simulator. The algorithms used in that study included Logistic Regression, Random Forest, and Decision Trees, with Logistic Regression used as a baseline model. Other related studies have shown that the Logistic Regression algorithm is capable of classifying fraudulent transactions well including credit card fraud, and insurance fraud, and it can even outperform complex algorithms [5,18,22,35,40]. Some of the advantages of Logistic Regression over traditional, rule-based approaches for financial fraud classification include being highly scalable. Hence it can be trained using large financial data, exhibits ease of interpretability, and can easily be deployed in industrial settings. However, it may be susceptible to overfitting, particularly when applied to inadequately preprocessed data.

The sigmoid function, $\sigma(x)$, is a crucial element in Logistic Regression and it maps any real-valued number into the range between 0 and 1. It is given by the equation

$$\sigma(x) = \frac{1}{1 + e^{-x}}, \tag{7}$$

where x is the input value to the function and e is the mathematical constant. The sigmoid function converts the linear combination of input features into a value that can be interpreted as the probability of belonging to a certain class in a binary classification problem inclusive of financial fraud classification. This implies that

$$P(Y = 1|X = x) = \frac{e\beta_0 + \beta_1 + X}{1 + e\beta_0 + \beta_1 + X}, \tag{8}$$

where Y takes on values between 0 and 1, $e\beta_0 + \beta_1 + X$ represents independent features and β_0, β_1 will yield different estimations of P.

The decision tree algorithm [25] uses the Gini index metric which is computationally less expensive compared to entropy, to determine the attributes that the tree should split on at every node. The Gini index is defined as

$$G(f) = \sum_{k=1}^{m} fk(1 - fk), \tag{9}$$

where fk represents the fraction of items labelled with k in the set and $\sum fk = 1$. The decision tree algorithms are highly effective in isolating fraudulent instances and often work together with rule-based systems and artificial neural networks.

The Random Forest classifier is a popular ensemble learning method that leverages the strengths of multiple decision trees for more accurate and robust predictions. It has become significant in identifying complex fraud patterns including money laundering [23,24,32]. This has made it to be one of the preferred choices in the financial industry for financial fraud detection, given that new unusual patterns keep on emerging in financial systems. Even though it can be superior in identifying complex patterns, it is computationally expensive when compared to the Logistic Regression algorithm [24].

Other algorithms consisting of KNN which is easy to set up due to its simplicity, and XGBoost which is well suited for large, complex datasets and offers superior performance are also used in financial fraud detection. However, KNN might not scale well with larger, high-dimensional data and XGBoost requires more expertise for fine-tuning [14,16].

The dataset used in the study [24] contained only a single fraud instance yet there is room for including more fraud scenarios especially when using a simulation platform and synthetic datasets for financial fraud detection research. On the other hand, our work explores the estimation of the total cost of fraud inclusive of hidden fraud in mobile money transactions, which is quite difficult to determine using real transaction data.

Lopez-Rojas et al. [28] introduce the analysis of fraud controls using the PaySim [28] financial simulator. PaySim generates synthetic financial data with a single fraud scenario. However, client attributes including email, national identification number, and phone number form part of a simulation model for studying sophisticated fraud scenarios. Besides, new fraud patterns keep on emerging and our study uses a simulation platform that has a detailed representation of the real mobile money ecosystem in Sub-Saharan Africa.

Fraud detection studies, especially for credit card fraud, include Altman [3], Sailusha et al. [39], and Thennakoon et al. [42] that are focused on fraud patterns found in the credit card ecosystems. Mobile money in the Sub-Saharan region presents unique fraud scenarios that in an event are not labeled in the real data need to be synthesized based on the unique processes of the real ecosystem before carrying out any fraud detection research. Our study addresses this gap and evaluates the effectiveness of existing fraud detection mechanisms while suggesting machine learning algorithms that can detect complex fraud patterns in mobile money transactions more accurately as compared to deterministic, rule-based approaches.

6   Conclusions

This study carried out experiments in the realm of mobile money financial services using the MoMTSim financial simulation platform. The experiments included an assessment of the statistical similarity of the synthetic data generated using MoMTSim to the real data. Estimation of the total cost of fraud in mobile money transactions. Financial fraud classification with a focus on comparing the performance of different machine learning algorithms for the task of financial fraud detection.

MoMTSim was employed to generate diverse mobile money transaction datasets, which were then assessed and found to statistically resemble the real data using the sum of squared errors (SSE) method. Common fraud scenarios were introduced into MoMTSim to enhance the synthetic datasets. We evaluated existing fraud controls by measuring the total cost of fraud and demonstrated that imposing a strict limit on mobile money transactions results in flagging many legitimate transactions as fraudulent, inconveniencing clients. On the other hand, a flexible limit on mobile money transfers allows many fraudulent transactions to go unnoticed in mobile money systems. This was evident in the classifier's performance, which exhibited a lower precision of 0.38 with a higher recall of 0.95 under strict control, whereas the recall decreased to 0.52 with an accompanying precision of 0.85 for the flexible transfer limit. These findings align with results observed in similar studies conducted in this research domain [27,28]. Service providers can minimize hidden fraud (false negatives) while reducing false alarms (false positives) by fine-tuning their fraud control mechanisms.

The efficacy of various machine learning classifiers, including Logistic Regression, Random Forest, Decision Trees, KNN, and XGBoost, in detecting financial fraud was assessed. This evaluation was conducted using a new dataset simulated with MoMTSim, featuring two fraudulent behaviors (refund fraud and fake credentials). Combining estimates for the total cost of fraud and model performance metrics for different machine learning classifiers, the Random Forest, and XGBoost algorithms emerged as efficient and consistent models for fraud detection. The Random Forest algorithm's ability to amalgamate predictions from numerous individual decision trees enables it to mitigate overfitting easily and enhance generalization, demonstrating the power of ensemble learning [24,32]. The XGBoost model incorporates L1 (Lasso) and L2 (Ridge) regularization to prevent overfitting by minimizing the model's complexity and distributing weights more evenly across all features [33,41,45]. However, simpler models, such as Logistic Regression, exhibited remarkable performance in fraud classification during the experiment and can thus be used in low-resource settings for fraud detection tasks.

Overall, the results from the experiments provide valuable insights for service providers to fine-tune their fraud controls. Additionally, the research community can further advance research in this domain to address anticipated fraudulent behaviors or emerging scenarios in the real ecosystem that may not be feasible to study using real transaction data. Therefore, automated fraud detection through computational methods is crucial in financial fraud prevention to stay a step ahead of criminals, unlike traditional deterministic, rule-based approaches that lack flexibility. This is primarily due to the high volume of transactions processed by mobile money systems, with new and unique behaviors making it challenging to identify suspicious activities using rule-based approaches.

The MoMTSim financial simulation platform, calibrated for mobile money transactions and synthetic datasets, will be made available to the research community to promote the study of new fraudulent behaviors and enhance privacy in synthetic data.

Acknowledgments. This research was made possible in part by the Digital Credit Observatory (DCO), a program of the Center for Effective Global Action (CEGA), with support from the Bill & Melinda Gates Foundation; JPMorgan Chase & Co.; and Google P.h.D Fellowship Program. Any views or opinions expressed herein are solely those of the authors listed, and may differ from the views and opinions expressed by any funder or its affiliates. Several financial institutions in the Sub-Saharan region provided expert opinions on the dynamics of the real financial ecosystem.

References

1. Akinyemi, B.E., Mushunje, A.: Determinants of mobile money technology adoption in rural areas of Africa. Cogent Soc. Sci. **6**(1), 1815963 (2020). https://doi.org/10.1080/23311886.2020.1815963
2. Ali, G., Ally Dida, M., Elikana Sam, A.: Evaluation of key security issues associated with mobile money systems in Uganda. Information **11**(6), 309 (2020). https://doi.org/10.3390/info11060309
3. Altman, E.: Synthesizing credit card transactions. In: Proceedings of the Second ACM International Conference on AI in Finance, pp. 1–9 (2021). https://doi.org/10.1145/3490354.3494378
4. Azamuke, D., Katarahweire, M., Bainomugisha, E.: Scenario-based synthetic dataset generation for mobile money transactions. In: Proceedings of the Federated Africa and Middle East Conference on Software Engineering, pp. 64–72 (2022). https://doi.org/10.1145/3531056.3542774
5. Bagga, S., Goyal, A., Gupta, N., Goyal, A.: Credit card fraud detection using pipeling and ensemble learning. Procedia Comput. Sci. **173**, 104–112 (2020). https://doi.org/10.1016/j.procs.2020.06.014
6. Botchey, F.E., Qin, Z., Hughes-Lartey, K.: Mobile money fraud prediction-a cross-case analysis on the efficiency of Support vector machines, Gradient boosted decision trees, and Naïve Bayes algorithms. Information **11**(8), 383 (2020). https://doi.org/10.3390/info11080383

7. Charbuty, B., Abdulazeez, A.: Classification based on decision tree algorithm for machine learning. J. Appl. Sci. Technol. Trends **2**(1), 20–28 (2021). https://doi.org/10.38094/jastt20165
8. Chawla, N.V., Bowyer, K.W., Hall, L.O., Kegelmeyer, W.P.: SMOTE: synthetic minority over-sampling technique. J. Artif. Intell. Res. **16**, 321–357 (2002). https://doi.org/10.1613/jair.953
9. Chen, T., Guestrin, C.: XGBoost: a scalable tree boosting system. In: Proceedings of the 22nd ACM SIGKDD International Conference on Knowledge Discovery and Data Mining, pp. 785–794 (2016). https://doi.org/10.1145/2939672.2939785
10. Chicco, D., Jurman, G.: An invitation to greater use of Matthews correlation coefficient in robotics and artificial intelligence. Front. Robot. AI **9**, 876814 (2022). https://doi.org/10.3389/frobt.2022.876814
11. Chicco, D., Tötsch, N., Jurman, G.: The Matthews correlation coefficient (MCC) is more reliable than balanced accuracy, bookmaker informedness, and markedness in two-class confusion matrix evaluation. BioData Mining **14**(1), 1–22 (2021). https://doi.org/10.1186/s13040-021-00244-z
12. Collier, N.: RePast: an extensible framework for agent simulation. Nat. Resour. Environ. Issues **8**(4) (2001)
13. Department of Computer Science, Makerere University: MoMTSim financial simulation platform (2023). version 1.0.0. www.github.com/aiinfinancegroup/MoMTSim
14. Faraji, Z.: A review of machine learning applications for credit card fraud detection with a case study. SEISENSE J. Manag. **5**(1), 49–59 (2022). https://doi.org/10.33215/sjom.v5i1.770
15. Guo, G., Wang, H., Bell, D., Bi, Y., Greer, K.: KNN model-based approach in classification. In: Meersman, R., Tari, Z., Schmidt, D.C. (eds.) On The Move to Meaningful Internet Systems 2003: CoopIS, DOA, and ODBASE. LNCS, vol. 2888, pp. 986–996. Springer, Heidelberg (2003). https://doi.org/10.1007/978-3-540-39964-3_62
16. Hajek, P., Abedin, M.Z., Sivarajah, U.: Fraud detection in mobile payment systems using an XGBoost-based framework. Inf. Syst. Front. 1–19 (2022). https://doi.org/10.1007/s10796-022-10346-6
17. Handelman, G.S., et al.: Peering into the black box of artificial intelligence: evaluation metrics of machine learning methods. Am. J. Roentgenol. **212**(1), 38–43 (2019). https://doi.org/10.2214/AJR.18.20224
18. Itoo, F., Meenakshi, Singh, S.: Comparison and analysis of Logistic regression, Naïve Bayes and KNN machine learning algorithms for credit card fraud detection. Int. J. Inf. Technol. **13**, 1503–1511 (2021). https://doi.org/10.1007/s41870-020-00430-y
19. Karpov, Y.G.: AnyLogic: a new generation professional simulation tool. In: VI International Congress on Mathematical Modeling, Nizni-Novgorog, Russia (2004)
20. Kolluri, J., Kotte, V.K., Phridviraj, M., Razia, S.: Reducing overfitting problem in machine learning using novel L1/4 regularization method. In: 2020 4th International Conference on Trends in Electronics and Informatics (ICOEI) (48184), pp. 934–938. IEEE (2020). https://doi.org/10.1109/ICOEI48184.2020.9142992
21. LaValley, M.P.: Logistic regression. Circulation **117**(18), 2395–2399 (2008)
22. Lokanan, M., Liu, S.: Predicting fraud victimization using classical machine learning. Entropy **23**(3), 300 (2021). https://doi.org/10.3390/e23030300
23. Lokanan, M.E.: Predicting money laundering using machine learning and artificial neural networks algorithms in banks. J. Appl. Secur. Res. 1–25 (2022). https://doi.org/10.1080/19361610.2022.2114744

24. Lokanan, M.E.: Predicting mobile money transaction fraud using machine learning algorithms. Appl. AI Lett. **4**(2), e85 (2023). https://doi.org/10.1002/ail2.85

25. Lokanan, M.E., Sharma, K.: Fraud prediction using machine learning: the case of investment advisors in Canada. Mach. Learn. Appl. **8**, 100269 (2022). https://doi.org/10.1016/j.mlwa.2022.100269

26. Lopez-Rojas, E., Elmir, A., Axelsson, S.: PaySim: a financial mobile money simulator for fraud detection. In: 28th European Modeling and Simulation Symposium, EMSS, Larnaca, pp. 249–255. Dime University of Genoa (2016)

27. Lopez-Rojas, E.A., Barneaud, C.: Advantages of the PaySim simulator for improving financial fraud controls. In: Arai, K., Bhatia, R., Kapoor, S. (eds.) CompCom 2019. Advances in Intelligent Systems and Computing, vol. 998, pp. 727–736. Springer, Cham (2019). https://doi.org/10.1007/978-3-030-22868-2_51

28. Lopez-Rojas, E.A., Axelsson, S., Baca, D.: Analysis of fraud controls using the PaySim financial simulator. Int. J. Simul. Process Model. **13**(4), 377–386 (2018). https://doi.org/10.1504/IJSPM.2018.093756

29. Luke, S., Cioffi-Revilla, C., Panait, L., Sullivan, K.: MASON: a new multi-agent simulation toolkit. In: Proceedings of the 2004 Swarmfest Workshop, Michigan, USA, vol. 8, pp. 316–327 (2004)

30. Luke, S., Cioffi-Revilla, C., Panait, L., Sullivan, K., Balan, G.: MASON: a multi-agent simulation environment. Simulation **81**(7), 517–527 (2005). https://doi.org/10.1177/0037549705058073

31. McNeish, D.M.: Using lasso for predictor selection and to assuage overfitting: a method long overlooked in behavioral sciences. Multivar. Behav. Res. **50**(5), 471–484 (2015). https://doi.org/10.1080/00273171.2015.1036965

32. Nami, S., Shajari, M.: Cost-sensitive payment card fraud detection based on dynamic Random forest and K-nearest neighbors. Expert Syst. Appl. **110**, 381–392 (2018). https://doi.org/10.1016/j.eswa.2018.06.011

33. Nti, I.K., Somanathan, A.R.: A scalable RF-XGBoost framework for financial fraud mitigation. IEEE Trans. Comput. Soc. Syst. (2022). https://doi.org/10.1109/TCSS.2022.3209827

34. Parmar, A., Katariya, R., Patel, V.: A review on Random forest: an ensemble classifier. In: Hemanth, J., Fernando, X., Lafata, P., Baig, Z. (eds.) ICICI 2018, pp. 758–763. Springer, Cham (2019). https://doi.org/10.1007/978-3-030-03146-6_86

35. Perols, J.: Financial statement fraud detection: an analysis of statistical and machine learning algorithms. Auditing A J. Pract. Theory **30**(2), 19–50 (2011). https://doi.org/10.1504/IJSPM.2018.093756

36. Powers, D.M.: Evaluation: from precision, recall and F-measure to ROC, informedness, markedness and correlation. J. Mach. Learn. Technol. **2**(1), 3763 (2011). https://doi.org/10.48550/arXiv.2010.16061

37. Sahin, Y., Bulkan, S., Duman, E.: A cost-sensitive decision tree approach for fraud detection. Expert Syst. Appl. **40**(15), 5916–5923 (2013). https://doi.org/10.1016/j.eswa.2013.05.021

38. Sahin, Y., Duman, E.: Detecting credit card fraud by ANN and logistic regression. In: 2011 International Symposium on Innovations in Intelligent Systems and Applications, pp. 315–319. IEEE (2011). https://doi.org/10.1109/INISTA.2011.5946108

39. Sailusha, R., Gnaneswar, V., Ramesh, R., Rao, G.R.: Credit card fraud detection using machine learning. In: 2020 4th International Conference on Intelligent Computing and Control Systems (ICICCS), pp. 1264–1270. IEEE (2020). https://doi.org/10.1109/ICICCS48265.2020.9121114

40. Sundarkumar, G.G., Ravi, V., Siddeshwar, V.: One-class support vector machine based undersampling: application to churn prediction and insurance fraud detection. In: 2015 IEEE International Conference on Computational Intelligence and Computing Research (ICCIC), pp. 1–7. IEEE (2015). https://doi.org/10.1109/ICCIC.2015.7435726

41. Tang, Q., et al.: Prediction of casing damage in unconsolidated sandstone reservoirs using machine learning algorithms. In: 2019 IEEE International Conference on Computation, Communication and Engineering (ICCCE), pp. 185–188. IEEE (2019). https://doi.org/10.1109/ICCCE48422.2019.9010785

42. Thennakoon, A., Bhagyani, C., Premadasa, S., Mihiranga, S., Kuruwitaarachchi, N.: Real-time credit card fraud detection using machine learning. In: 2019 9th International Conference on Cloud Computing, Data Science & Engineering (Confluence), pp. 488–493. IEEE (2019). https://doi.org/10.1109/CONFLUENCE.2019.8776942

43. Tian, Z., Xiao, J., Feng, H., Wei, Y.: Credit risk assessment based on Gradient boosting decision tree. Procedia Comput. Sci. **174**, 150–160 (2020). https://doi.org/10.1016/j.procs.2020.06.070

44. Tisue, S., Wilensky, U.: NetLogo: a simple environment for modeling complexity. In: International Conference on Complex Systems, vol. 21, pp. 16–21. Citeseer (2004)

45. Zhang, Y., Tong, J., Wang, Z., Gao, F.: Customer transaction fraud detection using XGBoost model. In: 2020 International Conference on Computer Engineering and Application (ICCEA), pp. 554–558. IEEE (2020). https://doi.org/10.1109/ICCEA50009.2020.00122

An Investigation and Analysis of Vulnerabilities Surrounding Cryptocurrencies and Blockchain Technology

Isabelle Heyl[1,2]($\boxtimes$) , Dewald Blaauw[1] , and Bruce Watson[1,2]

[1] Department of Information Science, Stellenbosch University,
Stellenbosch, South Africa
isabelle@nomail.co.za

[2] Centre for AI Research (CAIR), School for Data-Science and Computational
Thinking, Stellenbosch University, Stellenbosch, South Africa
https://www.sun.ac.za/english, https://www.cair.org.za

Abstract. As individuals search for a safe and alternative payment method, the popularity of cryptocurrencies and blockchain technology continues to grow. Despite their benefits, the increasing prevalence of these technologies also exposes them to a higher risk of encountering potential threats. These threats stem from the vulnerabilities and technical limitations within, which in turn, influences its potential to be adopted across different industries. Through a quantitative approach, this paper aims to detect and mitigate the vulnerabilities and technical limitations within cryptocurrencies and blockchain. Bitcoin was chosen because of its established reputation and widespread recognition in the market. A simulated environment provides the means to study distinct parameters within a Bitcoin network, particularly in the event of a double-spending and selfish mining attack. Based on the findings, the success rate and revenue of the attacker were primarily influenced by the network's stale block rate, the attacker's hash rate, and the double-spend value. Moreover, there is a significant difference in the parameters pre- and post-attacks. The block size has a considerable effect on the parameters in the network. As such, a single-objective problem is solved to determine what the optimal block size should be to minimize the block generation time and block delay time, seeking to reduce the overall latency and increase the throughput. It can be concluded that using the optimal block size can partially reduce the threat of double-spending attacks, but not eliminate it. It becomes evident that other mitigation schemes should be implemented to overcome these vulnerabilities.

Keywords: Cryptocurrencies · Blockchain · Double-spending · Selfish mining

T. G. Debelee et al. (Eds.): PanAfriConAI 2023, CCIS 2069, pp. 83–106, 2024.
https://doi.org/10.1007/978-3-031-57639-3_4

1   Introduction

In 2008, a group named Satoshi Nakamoto, introduced the first cryptocurrency in their paper "Bitcoin: A Peer-to-Peer Electronic Cash System" [1]. Shortly after this, the first Bitcoin platform was launched in 2009. The first block mined by the group yielded a reward of 50 bitcoins. This block is also known as the 'genesis block'. The first crypto transaction was between the group and Hal Finney, transacting 10 bitcoins. Satoshi Nakamoto shortly disappeared after mining approximately 1 million bitcoins and Gavin Andresen became the new face of Bitcoin.

The popularity of cryptocurrencies spiked in 2017, when one Bitcoin was approximately 14,800 USD and later reached its all-time high at the end of 2021 exceeding 61,000 USD [2]. However, Bitcoin saw its price crash by nearly 73 percent by the end of December 2022. Researchers argue that cryptocurrencies are too volatile, while others believe that their decentralized and secure nature may be beneficial.

The open-source code of Bitcoin allowed other developers to create different cryptocurrencies. As of February 2022, there are over 10,000 cryptocurrencies [2]. These digital currencies have removed hierarchical power, allowing people to make transactions with crypto coins electronically through peer-to-peer networks. Cryptocurrencies rely on digital data, digital transmission of data, and cryptographic methods to ensure authentic transactions [3]. The underlying technology of cryptocurrencies is blockchain technology, a decentralized distributed database system containing public ledgers [4].

Blockchain has gained attention from various domains, with many industries looking to adopt it. The main attraction of blockchain is its robustness, but many reports are emphasizing the vulnerabilities found within the technology [4]. Blockchain enables cryptocurrencies to achieve more secure, reliable, and transparent transactions, hence the importance of maintaining the technology to ensure it operates at its topmost form.

The vulnerabilities and technical limitations found within cryptocurrencies and blockchain need to be studied and understood to mitigate them for cryptocurrency and blockchain to be more widely accepted. This will not only benefit the crypto world, but also various industries wanting to adopt blockchain as a secure data storage platform.

Cryptocurrency and blockchain technology are becoming more popular each year. People are using it to decentralize their money and it has become an alternative payment method for big companies such as Burger King and Microsoft [5]. As it becomes more popular, more cyber risks arise. These risks stem from vulnerabilities within the blockchain, therefore it is important to understand and mitigate these vulnerabilities.

Dai et al., [6] among many other researchers argue that "blockchain is still in its exploratory stage and further development is needed for it to be optimized". Despite being considered as highly secure, blockchain has some limitations. The security challenges stem from its decentralized nature and self-organizing characteristics. These challenges include technical limitations such as performance and scalability and various attacks associated with blockchain.

While several research papers have discussed the challenges and vulnerabilities within cryptocurrencies and blockchain, such as scalability issues and attacks such as the 51% attack [6], few have focused on the impact that selfish mining and double-spending has on the network's parameters and how the technical limitations directly affect the security and overall performance.

The vulnerabilities within cryptocurrencies and blockchain, focusing on a Bitcoin network, will be investigated to see how they affect the technology. More specifically, how the technical limitations affect it. The purpose of this paper is to improve the quality to detect and mitigate these vulnerabilities.

2   Materials and Methods

This paper follows a quantitative approach to identify and analyze the technical limitations and vulnerabilities within a Bitcoin network. This is the preferred approach due to its ability to allow for a more scientific, controlled, and less biased outcome.

The simulations were run under different network circumstances. The output of these simulations was used to generate datasets. The simulation tool used in this method is NS-3, along with the Bitcoin simulator built by Arthur Gervais. Simulations are run individually and datasets are created accordingly. R was then used for data wrangling and to create graphs to identify patterns, relationships and trends within the data.

After the parameters were studied and hypotheses tested, a single-objective problem was solved to calculate the optimal block size. Python is the chosen language used for this.

2.1   Research Tools

The Bitcoin Simulator. This method is based on the Bitcoin Simulator, created by Arthur Gervais. Bitcoin Simulator is built on NS-3 - a discrete-event simulator. The purpose of the simulator is to analyze the potential impact that consensus parameters, network attributes, and changes to the protocol have on the scalability, security, and efficiency of Bitcoin [7]. The simulator has basic network properties and parameters such as block size, generating blocks, connectivity of nodes, block propagation, and other characteristics in the Bitcoin network [8]. Arthur Gervais wanted to create a realistic simulator, and to do so, he gathered real network data and integrated it into the simulator. More specifically, to collect the estimated values of block generation and block sizes, they crawled blockchain.info and used the bitcoin crawler to get an estimate of the number of nodes present in the network, as well as their geographic location. Coinscope also provided data regarding the node connectivity [7].

The Bitcoin Simulator provided the capability to adjust the parameters to study the technical limitations of Bitcoin, by changing parameters such as block size and the number of blocks. This simulator also provided the ability to execute a selfish miner test and perform double-spending attacks.

NS-3. NS-3 is a discrete-event network simulator [9]. It is built for the use of research and education. This free and open-source simulation environment is used for network research. The goal of NS-3 is to create an easy to use and debug and well documented form of simulation configuration in order to collect and analyze data. The software allows for realistic simulation models that can be used as a real time network emulator. NS-3 provides many internal libraries that can be used.

Waf. Waf is an open source and platform-independent automation tool written in Python [10]. Waf has no dependency on additional software or libraries. It does not require a code generator to enable efficient and extensible builds and the waf targets are defined as objects, separating defining targets from running commands [11].

RapidJSON. RapidJSON is a JSON parser and generator for C++. Its features include fast performance, self-contained and header-only, memory-friendly, and Unicode-friendly [12].

R. R is a highly extensible language used for statistical computing and graphics [13]. It can carry out various statistical analyses such as linear and nonlinear modelling, time series analysis, classification, clustering, and many other techniques, as well as graphical methods. RStudio (an integrated development environment) is used for R.

2.2   Simulation Parameters

The Bitcoin simulation has two main source code files, bitcoin-test.cc and selfish-miner-test.cc. Table 1 describes the input parameters and their corresponding default values.

Table 1. Input Parameters for Bitcoin Simulator [7]

Input Parameter	Description	Default Value
blockSize	This represents the block size (bytes). The default value matches the actual bitcoin block size obtained from blockchain.info. This value is fixed	−1
noBlocks	The number of generated blocks	100
nodes	The total number of nodes present in the network. This value should always be greater or equal to the number of miners	16
noMiners	The total number of miners in the network. This value and the miner's hash rates can only be changed in the minersHash array in the bitcoin-test.cc and selfish-miner-test.cc scripts	16
minConnections	The minConnectionsPerNode of the grid	−1
maxConnections	The maxConnectionsPerNode of the grid	−1
blockIntervalMinutes	The average time interval for block generation, in minutes	10
invTimeoutMins	The inv block timeout	−1
ud	The transaction value which is double spent	0
r	The stale block rate	0

2.3 Simulations

The simulations can be separated into three sections. For the first Sect. 2.3, the code from the bitcoin.cc file was used, where no attack was being tested. The different parameters of Bitcoin were first studied. For the second Sect. 2.3, the simulations were done on a network under the event of selfish mining and double-spending attacks and the last Sect. 2.4 involves the testing of the hypotheses.

The following table gives the parameters used in the datasets created from the simulations. Table 2 shows the parameters for datasets A, B, C, D, E, F, G, H, I, J, K, L, M.

Table 2. Parameters for datasets A, B, C, D, E, F, G, H, I, J, K, L, M

Parameter	Description
blockSize	The block size
noMiners	The number of miners in the network
T_{blocks}	Total blocks in simulation
att#	The attacker's hash rate
latency	The latency in the network, measured in ms
iteration	This represents one cycle of execution through a simulation
r	Determines the probability of a block becoming stale[1]
ud	The double-spend value[2]
$Total_{blocks}$	Total blocks in the network
s_{total}	The total amount of stale blocks
MBRT	Mean block receive time
MBPT	Mean block propagation time
r_s	The percentage of stale blocks in the network[3]
$Fork_L$	The longest fork in the chain
DS	The amount of successful double-spend attacks
Gen-B	The total number of blocks generated in the main chain by a miner
Gen-B (percent)	The percentage of a miner's generated blocks in the main chain
MBMB	The total number of mined blocks in the main chain
HM_{income}	The honest mining income, denoted by: $HM_{income} = Gen - B * (1 - r)$
ATT_{income}	The attacker income, denoted by: $ATT_{income} = (DS * ud) + MBMB$
ATT_{income} (percentage)	The attacker income fraction
HM-R	The relative revenue for honest miners: $HM - R = (HM_{income})/((HM_{income} + ATT_{income}))$
ATT-R	The relative revenue for attacker: $ATT - R = (ATT_{income})/((HM_{income} + ATT_{income}))$
t_s	The total time for an iteration, measured in seconds

[1]Larger r = higher probability of blocks becoming stale in the network. Smaller r = a lower probability of blocks becoming stale in the network.
[2]If $ud = 0$, the attacker follows the optimal strategy for selfish mining. If $ud \neq 0$, the attacker follows the optimal strategy for performing a double-spending attack of a transaction that has the same value as ud.
[3]Due to orphaned blocks or blocks that were not added to the main chain.

Simulating the Bitcoin Network Under No Attack. The data for the first dataset, A, are collected by running simulations based on the source file, bitcoin-test.cc. The relationship between bandwidth and block size is first studied. Increasing the number of miners in the network will distribute the mining power evenly across the network and can result in lower latency. An increase in miners will also increase the block confirmation times and consequently increase the throughput [14]. We therefore change the number of miners in the network to study the different effects it has on the rest of the parameters.

Simulating the Bitcoin Network with Adversaries' Present. The data were collected from running different simulations where an adversarial attacker is present in the network. The code in the selfish-miner-test.cc file was used. Firstly, the effect that the stale block rate, selfish mining and double-spending attacks have on the double-spending success was studied (A). Thereafter, the profitability of each attack was compared to honest mining (B). The emphasis was placed on parameters such as attacker's hash rate, latency, stale block rate, double-spending value, the number of double-spending successes, number of generated blocks, iterations, honest mining income and attacker income.

A. Double-spend success

Datasets B, C, D, E, and F are used for this section. The attacker in the simulations run for datasets B and C, follows the optimal strategy for selfish mining and in datasets D, E and F the attacker follows the optimal strategy for performing a double-spend attack. The number of double-spending successes are grouped into the following 4 categories: 0 Successes, 1–5 Successes, 6–10 Successes, and more than 10 Successes. Each simulation was done simulating over five iterations for each attack while keeping the noBlocks = 300 and changing the attacker's hash rate (0.1; 0.2; 0.3; 0.4; 0.5).

B. Profitability

The profitability of selfish mining and double-spending attacks was studied under different attacker's hash rates (att#) and increased stale block rates (r). A selfish miner in the network was first simulated, followed by an attacker following the optimal strategy for double spending with double-spending values of 10, 20, and 40.

2.4   Statistical Analysis

Statistical methods were used to determine the effect that the attacks had on the network. In other words, to determine whether there was a significant difference between the parameters before and after an attack. Dataset J contains the parameters and the corresponding values for a Bitcoin network where no adversary is present and dataset I contains the parameters and the corresponding values when an attacker is executing an attack. To ensure that the simulated networks for before the attacks and after the attacks were similar, the parameters for both networks were set as: noBlocks = 300, nodes = 16, noMiners = 16, and the miners hash rates the same for both before and after the attacks. From this, a new dataset, K, containing the parameters and their corresponding values from before attacks and after attacks was created.

The observations in dataset I are unrelated to dataset J, therefore the chosen statistical test was the independent samples T-test. This test was conducted to evaluate the potential difference in means from unrelated groups. To successfully test hypotheses, it is important that the data adheres to the assumptions of the independent samples T-test. The assumptions for performing an independent samples T-test are described as follows: the data should not contain any outliers, the data should be normally distributed, and the data should have equal variances. For the data that does not meet the requirements for an independent samples T-test, will be tested using the Mann-Whitney U test or the Welch t-test.

Hypotheses Tests. The mean block receives time hypothesis test.

Let: μ_1 = the population mean of the mean block receive time after the attack and μ_2 = the population mean of the mean block receive time before the attack

$$H_0 : \mu_1 = \mu_2$$
$$H_a : \mu_1 \neq \mu_2$$

Significance level: $\alpha = 0.05$

The mean block propagation time hypothesis test.

Let: μ_1 = the population mean of the mean block propagation time after the attack and μ_2 = the population mean of the mean block propagation time before the attack

$$H_0 : \mu_1 = \mu_2$$
$$H_a : \mu_1 \neq \mu_2$$

Significance level: $\alpha = 0.05$

The median block propagation time hypothesis test.

Let: μ_1 = the population mean of the median block propagation time after the attack and μ_2 = the population mean of the median block propagation time before the attack

$$H_0 : \mu_1 = \mu_2$$
$$H_a : \mu_1 \neq \mu_2$$

Significance level: $\alpha = 0.05$

The miners mean block propagation time hypothesis test.

Let: μ_1 = the population mean of the miners mean block propagation time after the attack μ_2 = the population mean of the miners mean block propagation time before the attack

$$H_0 : \mu_1 = \mu_2$$
$$H_a : \mu_1 \neq \mu_2$$

Significance level: $\alpha = 0.05$

The miners median block propagation time hypothesis test.

Let: μ_1 = the population mean of the miners' median block propagation time after the attack and μ_2 = the population mean of the miners' median block propagation time before the attack

$$H_0 : \mu_1 = \mu_2$$
$$H_a : \mu_1 \neq \mu_2$$

Significance level: $\alpha = 0.05$

The stale block rate hypothesis test.

$$H_0 : The\ two\ populations\ are\ equal$$
$$H_a : The\ two\ populations\ are\ not\ equal$$

Significance level: $\alpha = 0.05$

The longest fork in the network hypothesis test.

$$H_0 : The\ two\ populations\ are\ equal$$
$$H_a : The\ two\ populations\ are\ not\ equal$$

Significance level: $\alpha = 0.05$

The total bandwidth hypothesis test.

Let: μ_1 = the population mean of the total bandwidths after the attack and μ_2 = the population mean of the total bandwidths before the attack

$$H_0 : \mu_1 = \mu_2$$
$$H_a : \mu_1 \neq \mu_2$$

Significance level: $\alpha = 0.05$

2.5 Block Size Optimization

The goal was to find the block size with the least amount of delay. In other words, the block size that yields the lowest block propagation and transmission time and block generation times. An increased block size increases the propagation times and decreases the block generation time [15]. While on the other hand, smaller block sizes can decrease the propagation times and increase the generation time due to an influx of blocks and the network not being able to accommodate all transactions in the mempool.

The optimization problem was redefined as a single-objective optimization problem. The analysis was used to uncover the trade-off that exists between different objectives and to discover the solutions that determine the optimal balance between the objectives [16]. The single-objective problem was addressed using the particle swarm optimization algorithm (PSO).

Related studies have used algorithms such as the strength pareto evolutionary algorithm (SPEA), the whale optimization algorithm (WOA), particle swarm optimization (PSO), and multi-objective particle swarm optimization (MOPSO) [17]. The particle swarm optimization algorithm is employed because it offers both simplicity and accuracy. The parameters used in the blockchain network were set to be similar to Aygun and Arslan [15] and Reddy and Sharma [20].

Table 3 provides an overview on the notations used along with descriptions.

Table 3. Parameters for the calculation of the single-objective function

Parameter	Definition
T_g	The block generation time
S_{mp}	The total size of the mempool
blockSize	The size of each block
T_{oh}	The overall duration for overhead
T_t	The combined total of transactions in a block
T_{merkT}	Cumulative number of transactions in the Merkle Tree
T_{MTCT}	Merkle tree construction time
T_d	The block creation delay
h	The dept of the tree
T_p	Processing time
b_T	The bandwidth of a node
N_m	Number of miners connected
n	The number of nodes

The Multi-objective Problems. The multi-objective problems are defined by multiple equations. There are many parameters that need to be considered. Authors such as Aygun and Arslan [15], define the block generation time as T_g, and the block creation delay as T_d, denoted by Eq. 1 and Eq. 2.

$$T_g = \left(\frac{S_{mp}}{\text{blockSize}}\right) \cdot \left(T_{oh} + \frac{T_t}{T_{\text{merkT}}} \cdot T_{\text{MTCT}}\right) \tag{1}$$

$$T_d = h\left(T_p + \frac{\text{blockSize}}{b_T} \cdot N_m\right) \tag{2}$$

For the block generation time (T_g), the total size of the mempool is defined by S_{mp} and the size of each block is defined by blockSize. Therefore, the number of generated blocks can be calculated as S_{mp}/blockSize. In the equation, the block overhead timing is assumed to be unique with a S_{oh} size of data. The Merkle tree construction time, denoted by T_{MTCT} is calculated using the cumulative total of transactions contained within a block. Presume that the block size is modified to represent the total number of transactions included in the block. Each transaction size is represented as S_t. Therefore, the total number of transactions in a block is calculated by blockSize/S_t and the total number of transactions in the Merkle tree is denoted by T_{merkT}. The number of Merkle trees that are to be constructed can be calculated through T_t/T_{merkT}. Aygun and Arslan [15] sets the number of transactions to 10, keeping it fixed. The duration required for constructing a Merkle tree with 10 transactions is denoted by T_{MTCT} with an expected value of 0.02 s.

To calculate the block creation delay (T_d), the number of miners linked to the specific node (N_m) and the depth of the Merkle tree (h), was first calculated. These values are denoted by Eq. 3 and 4 [20].

$$N_m = (n-1) \cdot P_e \tag{3}$$

$$h = \log_\mu (n \cdot (N_m - 1) + 1) \tag{4}$$

The weighted sum method is used to create a single-objective function derived from the multi-objective problem. This function is denoted by O and can be seen below in Eq. 5 [15].

$$O = \boldsymbol{w} \cdot T_d + (1 - \boldsymbol{w}) \cdot T_g \tag{5}$$

The PSO algorithm was implemented using python version 3.10. Table 4 summarizes the values used in the equations and Table 5 gives the values used for the PSO.

Table 4. Input values for the PoW based blockchain network

Parameter	Value
n	10.000
N_m	8
T_{oh}	5 s
T_{merkT}	10
T_{MTCT}	0.02 s
S_{mp}	96000 Kb
RangeOfblockSize	[12, 1500] Kb
S_t	1.2 Kb
T_p	30 milliseconds
b_T	10 Mbps
μ	100 milliseconds
P_e	$8/(n-1)$

Table 5. Input value for PSO

Parameter	Value
c_1 acceleration coefficient	1.49445
c_2 acceleration coefficient	1.49445
weight	0.74
Population size	100

2.6 Mitigation Using the Optimal Block Size

The block size calculated from the optimization process will be employed to assess whether the optimal block size can effectively address Bitcoin's threat to double-spending attacks, hence mitigate the vulnerabilities.

Firstly, new datasets were created from running simulations with parameter values set as: number of blocks = 300, total nodes = 16, number of miners = 16, iteration = 1 and the attacker's hash rate = 0.3 for both datasets L and M, but using the optimal block size previously calculated for dataset M. The block sizes in dataset L were calculated using the default values in the simulation.

3 Results

3.1 Simulating the Bitcoin Network Under No Attack

The data in dataset A were grouped according to the blocksize and noMiners. The following parameters are fixed: noBlocks = 300, nodes = 48, blockIntervalMinutes = 10 min, and the number of block confirmations = 6. Table 6 provides an overview of the correlation values between block size and other parameters it is highly correlated to.

Table 6. Correlation between the block size and the parameters it is highly correlated to

Parameter	Block size
Mean block propagation time	0.96
Miners mean block propagation time	0.99
Stale block rate	0.75
Total bandwidth	1

3.2 Simulating the Bitcoin Network with Adversaries' Present

Double-Spend Success. The second set of simulations were done under the assumption that there is an adversary present in the network. The first objective was to determine how the stale block rate and double-spend value affects the double-spending success. Table 7 provides a summary of the double-spending success rates.

Table 7. Double-spending successes

Parameter	Dataset B value	Dataset C value	Dataset D value	Dataset E value	Dataset F value
r	0	1	0	0	0
ud	0	0	1	2	4
DS ≥ 1	68%	76%	80%	68%	76%

Profitability. Figures 1 and 2 provides the output of the data entries for datasets G and H where the attacker yielded a higher profit than honest miners.

```
#new dataset for I
I1 <- I %>%
  select(`att#`, r, ud, DS, Income_honestmining, Income_attacker)

#Empty array
i_array <- c()

#Code to see when the income for attackers > honest mining
for(i in 1:nrow(I1)) {
  if(I1$Income_attacker[i] > I1$Income_honestmining[i]){
    i_array[i] <- 1
  }else{
    i_array[i] <- 0
  }
}

#append array to dataset with new column 'prof'
I1$prof <- i_array

#Count how many times 1 and 0
x <- nrow(I1[I1$prof == 1,])
y <- nrow(I1[I1$prof == 0,])

#print out the percentages
print(paste0("Attacker income > Honest mining income: ", round((x/(x+y))*100,2), "%"))

## [1] "Attacker income > Honest mining income: 6.67%"

print(paste0("Attacker income < Honest mining income: ", round((y/(x+y))*100,2), "%"))

## [1] "Attacker income < Honest mining income: 93.33%"

#Print out the rows where the income for attackers > honest mining
I1 <- I1 %>%
  filter(I1$prof == 1)
I1

## # A tibble: 2 x 7
##   `att#`     r    ud    DS Income_honestmining Income_attacker  prof
##   <dbl> <dbl> <dbl> <dbl>               <dbl>           <dbl> <dbl>
## 1   0.5   0.3     0    15                104.             106     1
## 2   0.5   0.5     0    11                74.5              86     1
```

Fig. 1. Code to print out rows where selfish mining is more profitable than honest mining

```
#print out the percentages
print(paste0("Attacker income > Honest mining income: ", round((x/(x+y))*100,2), "%"))

## [1] "Attacker income > Honest mining income: 60%"

print(paste0("Attacker income < Honest mining income: ", round((y/(x+y))*100,2), "%"))

## [1] "Attacker income < Honest mining income: 40%"

#Print out the rows where the income for attackers > honest mining
J1 <- J1 %>%
  filter(J1$prof == 1)
J1

## # A tibble: 9 x 7
##   `att#`     r    ud    DS Income_honestmining Income_attacker  prof
##   <dbl> <dbl> <dbl> <dbl>               <dbl>           <dbl> <dbl>
## 1   0.4   0.2    10     8                90.4             136     1
## 2   0.5   0.2    10     8                106.             137     1
## 3   0.3   0.2    20     5                67.2             135     1
## 4   0.4   0.2    20     8                94.4             218     1
## 5   0.5   0.2    20    12                110.             326     1
## 6   0.2   0.2    40     1                46.4              47     1
## 7   0.3   0.2    40     2                62.4              94     1
## 8   0.4   0.2    40    10                96              470     1
## 9   0.5   0.2    40    16                132             755     1
```

Fig. 2. Output of all the rows where the attacker's income is more profitable than honest mining

3.3   Hypotheses

Table 8 provides the p-values for each parameter.

Table 8. P-values

Parameter	p-value
The mean block receive time hypothesis test	1.649e-13
The mean block propagation time hypothesis test	1.274e-07
The median block propagation time hypothesis test	1.171e-07
The miners mean block propagation time hypothesis test	5.767e-07
The miners median block propagation time hypothesis test	1.171e-07
The stale block rate hypothesis test	6.127e-08
The total bandwidth hypothesis test	0.002085

3.4   Block Optimization

We first evaluate the block generation time, T_g, and the block creation delay, T_d with respect to the different block sizes. The block delay time linearly increases as the block size grows, as illustrated in Fig. 3. On the other hand, as the block size grows larger, the block generation time decreases, as seen in Fig. 4. Here it is clear to the reader that there exists a trade-off between these parameters.

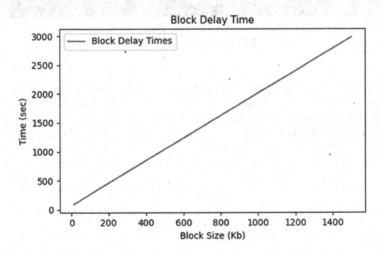

Fig. 3. Block delay time vs. block size (Kb)

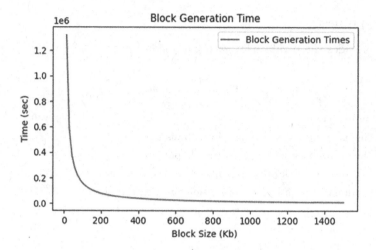

Fig. 4. Block generation time vs. block size (Kb)

Running the code given with 100 iterations and the parameters mentioned in Tables 4 and 5, the optimization process yields an optimal block size of 1198.26 Kb (Fig. 5).

```
Stopping search: Swarm best objective change less than 1e-08
Optimal Block Size: [1198.26217252] Kb
Objective Value: [6981.51615055]
```

Fig. 5. Output of PSO

While the weight increases, the block delay time decreases and the block generation time increases. From Figs. 6 and 7, it is clear that when the weight approximately reaches 0.8, the block delay time decreases and the block generation time increases. This further indicates the trade-off that happens between the two.

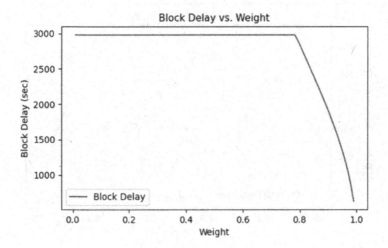

Fig. 6. Block delay time vs. weight

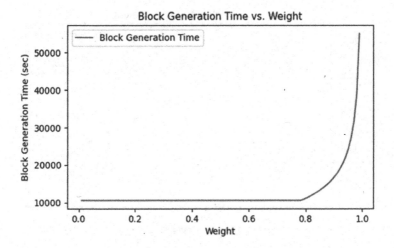

Fig. 7. Block generation time vs. weight

The optimal block size depends on the weight value and the component deemed more important. The turning point where the optimal block size changes happens when the weight value is 0.79, and can be seen in Fig. 8. For values lower than this threshold, larger block sizes are preferred, and for values above it, smaller block sizes are favored to achieve optimum time. Figure 9 shows how the objective value decreases as the weight increases.

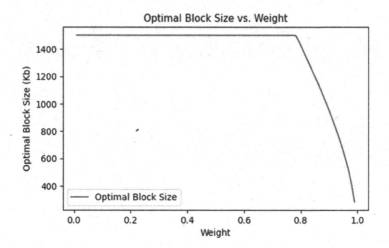

Fig. 8. The optimal block size vs. the weight

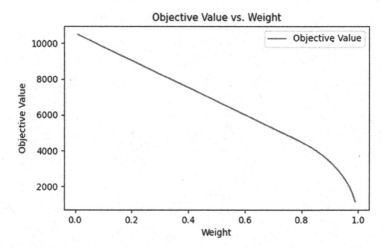

Fig. 9. The objective values vs. the weight

3.5 Mitigation

Here, the block size calculated from the optimization process will be employed to assess whether a block size of 1198.26 Kb (approximately 1.17 MB) can effectively address Bitcoin's threat to double-spending attacks, hence mitigate the vulnerabilities.

Firstly, new datasets are created from running simulations with parameter values set as: number of blocks = 300, total nodes = 16, number of miners = 16, iteration = 1 and the attacker's hash rate = 0.3 for both datasets L and M, but with the optimal block size set as 1198.26 Kb in dataset M. Table 9 summarizes and compares both datasets.

Table 9. A comparison of the parameters between dataset L and dataset M

Parameter	Dataset L value	Dataset M value
noBlocks	300	300
att(hash)	0.3	0.3
iterations	1	1
nodes	16	16
noMiners	16	16
blockSize (Kb)	513.76; 497.67; 496.68; 487.67; 499.50	1198.26

The block sizes in dataset L are calculated using the default values in the simulation. For dataset M, the block size is fixed to a value of 1198.26 Kb. The output of these simulations is summarized using the averages and is displayed in Table 10.

Table 10. The averages of each parameter

Parameter	Dataset L value	Dataset M value
blockSize	498.78	1198.26
MBRT	774.6184	796.5370
MNPT	364.29	345.0366
r_s	9.13650	9.13101
DS	3.6	3
b_T (Kbps)	12.64124	12.83176

4   Discussion and Conclusions

4.1   Simulating the Bitcoin Network Under No Attack

The reader was presented with an overview of how the Bitcoin Simulator works and the parameters associated with it. The first part of the simulations showed the Bitcoin network under no attack. This enabled the study of the relationships between the parameters. It was found that the block size is one of the key parameters that can affect the security measures, since it influences many other parameters. The block size was found to be highly correlated to the mean block propagation time, the miners mean block propagation time, the stale block rate, and the total bandwidth. These values can be seen in Table 6.

4.2 Simulating the Bitcoin Network with Adversaries' Present

Under the assumption of an adversary present in the network, the second set of simulations aimed to assess (i) how the double-spending success was influenced by the stale block rate and double-spend value and (ii) the profitability of each attack.

Comparing dataset B to C, it was found that even though there is no incentive for the attacker to perform a double-spending attack, the probability of having a successful double-spend attack is higher in dataset C. This can be due to the higher stale block rate in the network. Furthermore, the probability for a double-spend success is higher in dataset D than in dataset B. This can be attributed to the higher double-spend value and also because the attacker in dataset D is following the optimal strategy to perform a double-spending attack.

Comparing the network, C, where the attacker follows the optimal strategy for selfish mining with a higher stale block rate and the network, D, where the attacker follows the optimal strategy for double-spending, the double-spend success will still be more in a network such as in dataset D. Even though the stale block rate is higher in dataset C, the attacker in the network for dataset D has a higher double-spend value. Therefore, the attacker has more incentive to perform a double-spend attack.

Lastly, to compare datasets E and F, the stale block rate is set to 0 to determine whether an increased double-spending value will increase the double-spending successes. For an attacker with a higher double-spend value, the probability of having more double-spending successes is higher than in a network with an attacker with a lower double-spend value.

The second objective was to determine whether selfish mining and double-spending are profitable. The profitability of each attack is studied under the event of selfish mining and double-spending. It was found that for a selfish miner, their hash rate needed to be significantly high (0.5) along with a relatively high stale block rate ($r_s \geq 0.3$). It can be concluded that for selfish mining to be more profitable than honest mining, the attacker needs to own a large portion of the network's total hashing power (Fig. 1).

Under the circumstances when an attacker is trying to double-spend coins with the value of 10, 20, or 40, the attacker is only more likely to earn more than honest miners when the double-spend value is more than 10. The profitability of double-spending depends on the attacker's hash rate as well as the double-spend value. The higher the attacker's hash rate and the double-spend value, the higher their profit (Fig. 2).

4.3 Hypotheses

Hypothesis tests were used to determine whether the parameters have changed before and after an attack. It can be concluded that all the parameter values are significantly different before and after the attack, furthermore, showing the impact the parameters have on the security of blockchain.

4.4   Block Optimization

It was found that the block size is one of the key parameters that can affect the security measures, since it influences many other parameters. Related research [18,19] also provides sufficient evidence that block size affects the throughput and latency of the network. Therefore, the objective is to find the optimal block size. More specifically, findings from the research indicated that there exists a trade-off between the block generation time and the block delay time, therefore, the goal is to find the optimal block size which minimizes the block generation time and block delay time. Consequently, improving the throughput and reducing latency to maintain the security at its highest possible level. The block size optimization was done using the particle swarm optimization algorithm.

Based on the findings, for weights up until 0.79, larger block sizes are preferred to minimize the block generation time, while smaller blocks are preferred to minimize the block delay times for weight values less than 0.79. It can be concluded that the optimal block size depends on whether the researcher chooses to prioritize the block generation time or block delay time in the objective function.

4.5   Mitigation

In dataset L, the average attack success was 3.6, while in dataset M, it reduced to 3. This represents a reduction of approximately 16.67 percent in the attack success rate. Furthermore, the mean block propagation time and stale block rate reduced by 5.28 percent and 0.06 percent respectively. However, there was a slight increase of 2.83 percent in the mean block receive time. The bandwidth is expected to increase when the block size increases.

The mitigation process appears to have been partially successful. While the attack successes rate decreased by 16.67 percent, indicating an improvement in the security measures, there still exists a vulnerability to double-spending attacks.

4.6   Limitations

The Bitcoin Simulator provides a wide range of input parameters and ample room for experimentation, but there are limitations. The simulator does not allow users to change the transactions size distribution, the difficulty of the network, the transaction fee, and the throughput (TPS). These are important parameters and metrics that should be used to test the security of a blockchain network. Related studies have shown how these parameters affect the network and how they can be used to evaluate the security and performance of blockchain.

Another limitation is the limited number of blocks, nodes and miners in the simulated environment. The number of these parameters had to be reduced due to the constraints of the hardware used to run the application.

Furthermore, there is limited amount of documentation on the Bitcoin Simulator. Blockchain has only become a popular topic of discussion in recent years, therefore simulation tools have only recently been developed. These research tools are yet to be improved. Most blockchain simulators do not actively get updated and can result in outdated information [21]. There is a need for simulators that are continuously being improved to provide researchers with more accurate data.

4.7   Conclusion

It is possible to simulate a Bitcoin network to study the different parameters and technical limitations. Moreover, the effect that attacks can have on the network, specifically selfish mining and double-spending. However, it is possible to mitigate some of the limitations, using parameters such as the block size.

The stale block rate and double-spend value has an effect on the double-spending success. It can be concluded that a higher stale block rate and/ or higher double-spend value, can potentially increase an attackers double-spending success.

Furthermore, it was determined that selfish mining is not necessarily more profitable than honest mining. Selfish mining will only yield a higher profit if the attacker(s) own a significant amount of the network's total hashing power, which in actuality is not as straightforward. Moreover, double-spending will only be more profitable if the attacker(s) have a relatively high hash rate and well as double-spending a higher amount. As is the case with selfish mining, a higher hash rate will also benefit a double-spender.

Though selfish mining and double-spending may not be as profitable, it highly affects the parameters in the network. A change in parameters may increase the probability of another attack happening. The mitigations process focused on optimizing the block size parameter, since it was found that it highly affects the other parameters in the network. Moreover, it was also determined that block size highly affects the latency and throughput of the network, which we know influences the network's security.

The optimization process was successful and the trade-off that exists between the block generation time and block delay time was successfully minimized. The optimal block size will depend on which parameter is deemed more important than the other. The choice between prioritizing either block generation time and block delay time solely depend on the type of blockchain and cryptocurrency. From the mitigation process, it is evident that using the optimal block size can mitigate the vulnerabilities in the network but cannot be used as the only mitigation.

4.8   Future Research

It was firstly noted that there exists a gap in simulation-based research for blockchain networks. To support the advancement of blockchain systems research, the development of additional simulation platforms is required. This will allow researchers to address issues within blockchain in a cost-effective and accessible manner.

As of today, there does not exist a blockchain specific simulation platform where all parameters are included (such as TPS, latency, transaction fee and so on). There are many research papers comparing the different blockchain simulators built on other platforms such as NS-3. Nevertheless, they share a limitation concerning the limited number of parameters available for modification.

This paper seeks to address the technical limitations and vulnerabilities within cryptocurrencies and blockchain technology and a few key issues were discussed and mitigated, but there are many other limitations and vulnerabilities that each need to be addressed in a different manner beyond simulations. The scalability limitation within blockchain calls for a comprehensive approach in a larger real-world setting, as relying solely on a small, simulated network environment would be inadequate.

Future research can be done on the technical limitations of blockchain beyond cryptocurrency applications. One of the biggest limitations of blockchain is that it is not yet widely accepted. This issue can be addressed if more research is done within different sectors such as finance or healthcare. Furthermore, this will assist in educating a broader audience about the diverse array of applications that blockchain technology offers beyond merely cryptocurrencies.

5   Datasets

All datasets are available and can be accessed through the authors of this paper.

References

1. Chohan, U.W.: The double spending problem and cryptocurrencies. SSRN Electron. J. (2017). https://doi.org/10.2139/ssrn.3090174, Accessed 10 Jul 2023
2. de Best, R.: Number of crypto coins 2013–2022. Statista (2022). www.statista.com/statistics/863917/number-crypto-coins-tokens/, Accessed 11 Oct 2022
3. Farell, R.: An analysis of the cryptocurrency industry. Thesis, Wharton Research Scholars (2015). www.repository.upenn.edu/handle/20.500.14332/49177, Accessed 10 Oct 2022
4. Ghosh, A., Gupta, S., Dua, A., Kumar, N.: Security of Cryptocurrencies in blockchain technology: state-of-art, challenges and future prospects. J. Netw. Comput. Appl. **163**, 102635 (2020). https://doi.org/10.1016/j.jnca.2020.102635, www.sciencedirect.com/science/article/pii/S1084804520301090
5. Beigel, O.: Who accepts bitcoins in 2022? (2022). www.99bitcoins.com/bitcoin/who-accepts/, Accessed 02 Feb 2023
6. Dai, F., Shi, Y., Meng, N., Wei, L., Ye, Z.: From Bitcoin to cybersecurity: a comparative study of blockchain application and security issues. In: 2017 4th International Conference on Systems and Informatics (ICSAI), pp. 975–979 (2017). https://doi.org/10.1109/ICSAI.2017.8248427, Accessed 02 Feb 2023
7. Gervais, A.: Arthurgervais/Bitcoin-Simulation. www.github.com/arthurgervais/Bitcoin-Simulator (2016), Accessed 21 Apr 2023
8. Mufleh, A.: Bitcoin eclipse attack - statistic analysis on selfish mining and double-spending attack. Published master's thesis, Johannes Kepler University Linz, Austria (2019). www.epub.jku.at/download/pdf/3853668.pdf, Accessed 23 Mar 2023

9. NS-3 Project Team: What is NS-3? (n.d.). www.nsnam.org/about/what-is-ns-3/, Accessed 20 Mar 2023
10. Benson, D.: WAF: an excellent build automation tool, open source for You (2017). www.opensourceforu.com/2017/02/waf-excellent-build-automation-tool/, Accessed 20 Mar 2023
11. The Waf Book. www.waf.io/book/, Accessed 20 Mar 2023
12. RapidJSON Documentation. (2015). www.rapidjson.org/, Accessed 20 Mar 2023
13. R Core Team: What is R? (n.d.). www.r-project.org/about.html, Accessed 20 Mar 2023
14. Goswami, S.: Scalability analysis of blockchains through blockchain simulation. University of Nevada, Las Vegas (2017). Published master's thesis. www.dx.doi.org/10.34917/10985898
15. Aygün, B., Arslan, H.: Block size optimization for POW consensus algorithm based blockchain applications by using whale optimization algorithm. Turk. J. Electr. Eng. Comput. Sci. **30**(2), 406–419 (2022). https://doi.org/10.3906/elk-2105-217
16. Sela, L.: Lecture notes: multi-objective optimization (2020). www.digitalcommons.usu.edu/cgi/viewcontent.cgi?article=1091&context=ecstatic_all, Accessed 24 May 2023
17. Singh, N., Vardhan, M.: Computing optimal block size for blockchain-based applications with contradictory objectives. Procedia Comput. Sci. **171**, 1389–1398 (2020). https://doi.org/10.1016/j.procs.2020.04.149
18. Wilhelmi, F., Barrachina-Muñoz, S., Dini, P.: End-to-end latency analysis and optimal block size of proof-of-work blockchain applications. IEEE Commun. Lett. **26**(10), 2332–2335 (2022). https://doi.org/10.1109/LCOMM.2022.3194561
19. Croman, K., et al.: On scaling decentralized blockchains (A Position Paper). In: 3rd Workshop on Bitcoin and Blockchain Research, Barbados, February 2016. www.researchgate.net/publication/292782219_On_Scaling_Decentralized_Blockchains_A_Position_Paper, Accessed 27 Oct 2023
20. Reddy, B.S., Sharma, G.V.V.: Optimal transaction throughput in proof-of-work based blockchain networks. In: The 3rd Annual Decentralized Conference on Blockchain and Cryptocurrency, LNCS, vol. 28, no. 1, pp. 6. Springer, Heidelberg (2019). https://doi.org/10.3390/proceedings2019028006
21. Paulavicius, R., Grigaitis, S., Filatovas, E.: A systematic review and empirical analysis of blockchain simulators. IEEE Access **9**, 38010–38028 (2021). https://doi.org/10.1109/ACCESS.2021.3063324

Towards a Supervised Machine Learning Algorithm for Cyberattacks Detection and Prevention in a Smart Grid Cybersecurity System

Takudzwa Vincent Banda[1]([✉]) [iD], Dewald Blaauw[1,2] [iD],
and Bruce W. Watson[1] [iD]

[1] Stellenbosch University, Center for AI Research (CAIR), School of Data Science and Computational Thinking University, Stellenbosch, South Africa
22588930@sun.ac.za
[2] Department of Information Science, Stellenbosch University, Stellenbosch, South Africa

Abstract. Critical infrastructure cyberattacks have become a significant threat to national security worldwide. Adversaries exploit vulnerabilities in communication networks, technologies, and protocols of smart grid control systems network to gain access and control of power grids, causing blackouts. Despite the need to safeguard the reliable and stable operation of the grid against cyberattacks, simultaneously detecting and preventing attacks presents a significant challenge. To address this, a Kali Linux machine was connected to a smart grid control system network emulated in GNS3 to perform common cyberattacks. Wireshark was then deployed to capture network traffic for machine learning. Aiming to improve the detection and prevention of cyberattacks the study proposed a dual-tasked ensemble supervised machine learning model, a combination of Neural Network and Extreme Gradient Boosting, that had an average accuracy of 99.60% and detection rate of 99.48%. The first task of the model distinguishes between normal state and cyberattack modes of operation. The second task prevents suspicious packets from reaching the network destination devices. Leveraging the Power-Shell Script, the model dynamically applies packet filtering firewall rules based on its predictions. The proposed model was tested on new data, producing an accuracy of 99.19% and a detection rate of 98.95%. Furthermore, the model's performance was compared to existing proposed cyberattack detection models. Thus, the proposed model, with its function as a firewall, enhances the overall security capabilities of the smart grid and significantly mitigates potential cyberattacks.

Keywords: Smart Grid · Supervised Machine Learning · Control System · Cybersecurity · Cyberattacks · Neural Network · XGBoost · Firewall · Network Traffic

T. G. Debelee et al. (Eds.): PanAfriConAI 2023, CCIS 2069, pp. 107–128, 2024.
https://doi.org/10.1007/978-3-031-57639-3_5

1   Introduction

In recent years, there has been a growing global concern surrounding cyberattacks targeting critical infrastructure, with a particular focus on control systems [1]. Notable incidents, such as the 2010 Stuxnet malware attack on Iranian nuclear centrifuges [2], have highlighted the vulnerabilities of control systems within critical infrastructure. This attack aimed to disrupt Iranian enrichment by compromising programmable logic controllers (PLCs) and industrial machinery within the facility. Additionally, in December 2015 and 2016, Russia orchestrated False Data Injection cyberattacks on Ukraine's power grid through SCADA substations [3], resulting in power outages affecting approximately 225,000 customers.

In 2022, reports emerged regarding cyberattacks on the Grand Ethiopian Renaissance Dam [4], exposing significant cybersecurity challenges in Ethiopia linked to limited computer technology, insufficient cyber awareness, and the absence of effective cybersecurity policies in control centers. Moreover, in 2019, Johannesburg's City Power in South Africa fell victim to a ransomware incident, causing disruptions in a substantial portion of the utility's applications and networks, potentially impacting around 250,000 customers [5]. These incidents underscore the vulnerability of critical infrastructures to cyber threats and the urgent need for enhanced protection measures.

In the contemporary digital era, a pivotal element of critical infrastructure is the smart grid [6]. This advanced intricate electrical and communication network leverages digital innovations to revolutionize power generation, distribution, and consumption procedures. It facilitates two-way communication between utility providers and consumers through a sophisticated array of intelligent devices. The smart grid extends its influence over diverse domains, encompassing power generation, transmission, distribution, customer services, market dynamics, service providers, and operational aspects, all benefiting from advancements in information and communication technologies.

Nonetheless, the smart grid is not immune to the risk of cyberattacks. Its intricate network of devices, communication technologies, and protocols introduces vulnerabilities in various forms. These vulnerabilities often arise from the use of outdated technologies, reliance on aging infrastructure, the presence of zero-day vulnerabilities in certain technologies, the deployment of fragile digital communication networks, protocols, and the influence of human factors [7]. Control systems that underpin critical infrastructure are attractive targets for adversaries due to their central role in processing vast volumes of sensitive data. They oversee the monitoring and management of smart grid control centers, substations, and the broader grid network, underscoring the urgent need for robust cybersecurity measures.

Machine learning plays a pivotal role in smart grid cybersecurity by detecting and preventing cyber threats. Common machine learning techniques deployed in smart grid cybersecurity include anomaly detection, intrusion detection systems (IDS), malware detection, threat intelligence and analysis, encryption and data security, access control, and predictive maintenance [8].

Despite the application of various machine learning techniques in smart grid cybersecurity, challenges persist. Current models detect cyberattacks and often require additional tools or human intervention to prevent attacks. This sequential approach can allow malicious traffic to penetrate the network before mitigation. In contrast, this paper will challenge the current approach by proposing an advanced supervised machine learning algorithm that simultaneously detects and prevents cyberattacks by providing real-time firewall rules, representing a significant advancement in smart grid cybersecurity.

To achieve this objective, the paper is guided by the following central question: How can machine learning techniques be integrated with firewall rules to simultaneously detect and prevent cyberattacks in the control systems of the smart grid and smart grid infrastructure at large?

The remainder of the paper unfolds as follows: In Sect. 2, a comprehensive literature review delves into existing research on smart grid cybersecurity and machine learning. Section 3 outlines the research design, detailing the data collection and preprocessing methods, and introduces the proposed advanced supervised machine learning model. Section 4 presents the results from the experiments and analyses such results obtained. Section 5 provides study discussions and implications. Section 6 presents our concluding observations, summarizing the key takeaways from the study. Section 7 offers recommendations for future research and explores potential avenues to further advance the field of smart grid cybersecurity. Lastly, will provide acknowledgments.

2   Literature Review

This section provides a review of existing literature related to the application of machine learning to protect smart grid control systems and previously proposed machine techniques by other authors to enhance the security of these control systems.

2.1   Machine Learning Application in Smart Grid Control Systems

Smart grid control systems are the foundation of modern electricity distribution, managing power generation, transmission, and distribution [9]. They create data-driven environments, prioritizing automation and control for rapid adaptation to changing conditions [9]. Various devices, including sensors, continually gather data on grid states such as voltage, current, frequency, and environmental parameters. A sophisticated communication network transmits this critical data to central control centers through substations [10]. Despite their complexity, control systems face cybersecurity challenges, including threats like Distributed Denial of Service (DDoS), Denial of Service (DoS), False Data Injection Attack (FDIA), and Man-in-the-Middle Attacks (MitMA) [11].

Machine learning (ML) is pivotal for enhancing control system cybersecurity, offering diverse mechanisms to protect data, and networks and combat ransomware threats [8]. Advanced ML algorithms reinforce data protection through encryption, managing encryption keys dynamically, and detecting unauthorized access [12]. ML empowers intrusion detection by continually analyzing network traffic patterns and identifying irregularities with anomaly detection [13]. ML-enhanced security measures can identify indicators of compromise (IOCs) in real-time, enabling swift action against ransomware attacks [14]. This comprehensive use of ML ensures data integrity, secures communication channels, and actively prevents cyber threats from infiltrating critical systems.

Consequently, this paper will focus on the use of machine learning to analyze network packets transmitted within the grid control systems and those exchanged between the control system network and external sources to discern and isolate foreign packets.

2.2 Related Work

Authors in [15] proposed a novel hybrid machine learning approach that incorporated both supervised and unsupervised models to detect spoofing attacks within a smart grid SCADA system. Their research, utilizing the IEEE 14-bus system simulation model, demonstrated the efficacy of ensemble models in achieving a 73% detection rate, surpassing the performance of individual supervised and unsupervised models. However, the study recognized the challenges associated with the computational complexity of implementing intricate architectures and emphasized the importance of addressing real-world scenarios.

In another study by [16], the focus was on theft identification and FDIA detection in a SCADA server, employing Random Forest (RF). By utilizing the multi-class Theft Detection Dataset (TDD2022), the researchers implemented various models, including RF, KNN, DT, and ANN. Notably, Random Forest (RF) outperformed the other models, achieving a remarkable 99.4% accuracy and a detection rate of 99.2%. Nevertheless, the study suggested that improvements could be made in feature selection to enhance recognition and inferencing overhead.

Furthermore, authors in [17] proposed an Extremely Randomised Trees algorithm designed to detect stealthy False Data Injection Attacks across the entire smart grid system. Their research involved the utilization of simulation models for the IEEE 57-bus and IEEE 118-bus systems. Additionally, Kernel Principal Component Analysis (KPCA) was applied for dimensionality reduction. The algorithm exhibited a high accuracy of 97% in detecting stealthy FDIA. However, it is important to note that the study primarily focused on FDIA and did not encompass other categories of cyber threats within the smart grid.

Authors in [18] introduced the Probabilistic Bayesian Regression (PBR) algorithm, which was dedicated to detecting various classes of attacks within a smart grid control system. Leveraging a DC power flow model, they achieved an impressive 97% accuracy in classifying attacks. However, the research was limited by its concentration on only two types of attacks, leaving room for further exploration into algorithm performance with a broader range of attack types.

In a proactive approach, contributing to the Intrusion Prevention Systems (IPS) for smart grids, authors in [19] designed a learning firewall model with

CART dependency. This system dynamically adjusted by crafting preventive rules and reducing false alarms in automated control operations. Notably, their approach relied on a zero false-positive metric for decision-making, effectively creating a machine learning-based firewall system. The model was evaluated using the KDD Cup 99 dataset and achieved an accuracy rate of 96%.

3 Materials and Method

This section describes the methodology for developing a cybersecurity model for smart grid control systems using machine learning-based analysis of network packet traffic. It covers tool selection, experimental setup, data preprocessing, model development, training, testing, hyperparameter settings, and model evaluation for detection tasks. Additionally, it unpacks how the integration of firewall rules with the model's decisions to proactively prevent cyberattacks was implemented.

3.1 Materials

For a comprehensive study, a carefully selected set of tools and technologies was employed. Graphical Network Simulator-3 (GNS3) served as the primary open-source platform for emulating the control system network. In conjunction, VMWare Workstation Player was utilized for hosting GNS3 because it offers faster performance compared to alternatives [20]. Kali Linux was chosen for simulating cyberattacks because it provides a diverse range of penetration tools [21]. Wireshark captured vital network packets for subsequent analysis, model training, and testing [22]. Python, along with libraries like NumPy, SciPy, SciKit-Learn, Pandas, Matplotlib, and Seaborn, handled machine learning tasks and data visualization. The Subprocess Module and PowerShell Script executed firewall rules, and Joblib managed efficient model loading and saving. The logging module recorded crucial information about network packets, and Jupyter Notebook provided an interactive coding environment for machine learning tasks and data visualization [23].

3.2 Experiment Design

In the experiment design (see Fig. 1), a smart grid control system was emulated in GNS3 [24]. For a comprehensive guide on establishing the smart grid control system network topology, one can refer to the work of [25]. This approach leverages the unique attributes of the smart grid context to validate and demonstrate real-world scenarios.

The network architecture included the control center for centralized monitoring and management, an automated substation for grid status regulation and data collection, and a field area that encompasses the physical infrastructure for power distribution and transmission.

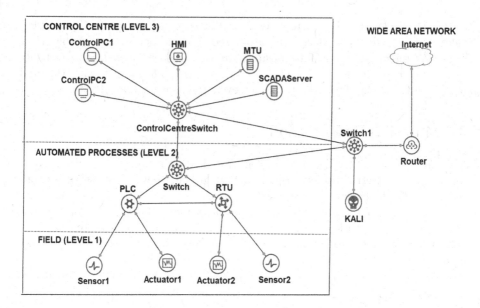

Fig. 1. Emulated control system network in GNS3

Devices within the control center such as Control PCs, control system servers, and Master Telemetry Units (MTUs) consistently interacted with Programmable Logic Controllers (PLCs) and Remote Terminal Units (RTUs) in the automated substation through information and communication technologies (ICT). Sensors located in the field area played a critical role in continuously capturing essential grid status data.

This data automatically underwent processing within the substation before being presented in the control center for analysis and control, highlighting the data-driven nature and harnessing digital innovations of control systems. The network was also connected to the external network (Internet) via a router for data exchange with external entities. Furthermore, to emulate potential threats, a network intruder Kali Linux, was connected to the network main switch as an entry point to perform cyberattacks.

3.3   Data Collection

The data collection phase involved the capturing of real-time control system network packet traffic in its normal state and during attacks using Wireshark (see Fig. 2).

All captured packets were saved in a comma-separated value (CSV) file. Captured packet 12 features included Time, IP addresses, MAC addresses, ports, time to live, flags, ICMP code, and ICMP type. IP addresses are numerical labels facilitating device communication on a network; MAC addresses are unique identifiers at the hardware level for network interfaces; ports are logical end-points directing network traffic to specific applications; Time to Live (TTL) is a value in packet headers determining the maximum network hops; flags are conveying control and status information in packet headers; ICMP code provides specific network condition details within ICMP packets; and ICMP type is a category of the purpose or function of ICMP messages in network communication.

The normal network packets traffic category represented normal network performance without any significant drops or unacceptable changes. A total of 84 scenarios were implemented and repeated twice at different intervals to collect data. The scenarios involved activities such as writing files in sensors to simulate data collected from powerlines and transformers, transferring data from sensors to RTUs and PLCs, performing echo ping requests and replies between all network devices, uploading, and delivering files between devices and servers, and accessing data on servers using workstation control PCs.

Time	Src_id	Dst_id	Src_mac	Dst_mac	Protocol	Src_port	Dst_port
0.001	10.10.10.12	10.10.10.21	da:e5:bb:12:23:b4	9e:cc:7e:c5:73:1e	FTP	21	37738
0.000	10.10.10.21	10.10.10.12	9e:cc:7e:c5:73:1e	da:e5:bb:12:23:b4	FTP	37738	21
0.000	10.10.10.12	10.10.10.21	da:e5:bb:12:23:b4	9e:cc:7e:c5:73:1e	FTP	21	37738
0.000	10.10.10.21	10.10.10.12	9e:cc:7e:c5:73:1e	da:e5:bb:12:23:b4	FTP	37738	21
1.044	10.10.10.12	10.10.10.21	da:e5:bb:12:23:b4	9e:cc:7e:c5:73:1e	FTP	21	37738
0.001	10.10.10.12	10.10.10.21	da:e5:bb:12:23:b4	9e:cc:7e:c5:73:1e	FTP	21	54348
0.000	10.10.10.21	10.10.10.12	9e:cc:7e:c5:73:1e	da:e5:bb:12:23:b4	FTP	54348	21
0.000	10.10.10.12	10.10.10.21	da:e5:bb:12:23:b4	9e:cc:7e:c5:73:1e	FTP	21	54348
0.000	10.10.10.21	10.10.10.12	9e:cc:7e:c5:73:1e	da:e5:bb:12:23:b4	FTP	54348	21

Fig. 2. Wireshark capturing network traffic in real-time. In this scenario, Wireshark was capturing file transfer network packets from sensors to RTU and PLC devices in the substation under normal conditions.

The suspicious network packets traffic category included all the packets' traffic from attacks on network devices aimed to affect the network's normal performance. Therefore, all events that were intended to drop or change the behavior of the network's normal operations were captured. A variety of attacks summarized in Table 1 were performed on the network using Kali.

Table 1. Summary of implemented attacks on the network [26–28].

Attack	Sub-Attack	Tools	Goal of the Attack
DoS	TCP-SYN flood Spoofed IP address Ping of death DHCP Starvation DNS Spoofing	Hping3 Ping Yersinia	Overwhelm and disrupt target all network devices from a single source
DDoS	UDP flood ICMP flood Random Source Attack PUSH-ACK flood, TCP-RST flood HTTP flood	Hping3 Xerxes	Flood targeted devices especially servers with high traffic from different sources
MitMA	ARP Poisoning NDP Poisoning Port Stealing	Ettercap	Intercept and alter communication between devices in different sections
FDIA	Manipulate sensor data. Manipulated packets	Man-in-the-middle techniques Scapy	Inject false data into the system and sensors

There were two resulting datasets: a training and testing dataset and a previously unseen dataset (new data) for model development. The training and testing dataset, before the split, featured 12 input features and a total of 63,845 data points, and previously unseen data featured 27,919 data points. These datasets with "Normal" and "Suspicious" labels laid the groundwork for robust model training and evaluation. (see Fig. 3) depicts the snippet of the dataset.

	Time	Src_ip	Dst_ip	Src_mac	Dst_mac	Protocol	Src_port	Dst_port	T_to_live	Flags	ICMP_type	ICMP_code	Traffic_Class
0	0.021	10.10.10.4	10.10.10.9	0c:85:46:d8:00:00	c2:71:a5:be:36:82	TCP	42974	21	64	2	0	0	Normal
1	0.000	10.10.10.9	10.10.10.4	c2:71:a5:be:36:82	0c:85:46:d8:00:00	TCP	21	42974	64	18	0	0	Normal
2	0.002	10.10.10.4	10.10.10.9	0c:85:46:d8:00:00	c2:71:a5:be:36:82	TCP	42974	21	64	16	0	0	Normal
3	0.001	10.10.10.9	10.10.10.4	c2:71:a5:be:36:82	0c:85:46:d8:00:00	FTP	21	42974	64	24	0	0	Normal
4	0.010	10.10.10.4	10.10.10.9	0c:85:46:d8:00:00	c2:71:a5:be:36:82	TCP	42974	21	64	16	0	0	Normal
5	0.000	10.10.10.4	10.10.10.9	0c:85:46:d8:00:00	c2:71:a5:be:36:82	FTP	42974	21	64	24	0	0	Normal
6	0.000	10.10.10.9	10.10.10.4	c2:71:a5:be:36:82	0c:85:46:d8:00:00	TCP	21	42974	64	16	0	0	Normal
7	0.000	10.10.10.9	10.10.10.4	c2:71:a5:be:36:82	0c:85:46:d8:00:00	FTP	21	42974	64	24	0	0	Normal
8	0.018	10.10.10.4	10.10.10.9	0c:85:46:d8:00:00	c2:71:a5:be:36:82	FTP	42974	21	64	24	0	0	Normal
9	0.043	10.10.10.9	10.10.10.4	c2:71:a5:be:36:82	0c:85:46:d8:00:00	TCP	21	42974	64	16	0	0	Normal

Fig. 3. Sample of the generated datasets displayed using Python Pandas with 12 input features (Time, Source and Destination IP address, Source and Destination MAC address, Protocol, Source and Destination Ports, Time to live, Flags, ICMP type, and ICMP code, and output feature (Traffic Class)).

3.4 Data Preprocessing

Data preprocessing is a critical initial step in addressing data science challenges, particularly when preparing data for machine learning models. Raw data often contains various issues such as noise, inconsistencies, and missing values [29].

These issues can hinder the accuracy and reliability of machine learning models. In the context of this study, dealing with network packet data characterized by numerous categorical and discrete variables posed additional challenges related to dimensionality, sparsity, encoding, and interpretability. Therefore, it was necessary to implement data preprocessing.

1. Data Cleaning

One significant aspect of data pre-processing is handling missing or NaN (Not a Number) values. In the datasets, missing values were prevalent due to the nature of the collected data. For instance, ICMP protocols often lacked Source Port and Destination Port values, while other protocols like TCP, FTP, and HTTP lacked ICMP type and ICMP code information. While common approaches are to remove entire rows containing missing values or replace them with average values, this paper opted to retain rows with missing values. Instead, these missing values were systematically replaced with zeros, aligning with the dataset's requirements and the specific machine learning models used.

2. Data Transformation

Data transformation was crucial to prepare the datasets for analysis by machine learning algorithms. Transformation techniques included encoding categorical variables into numerical representations and normalizing numerical features to improve predictive accuracy.

i. *Encoding the Target Variables*: Categorical target classes were encoded into numeric binary variables, enhancing model performance without information loss.

ii. *Encoding X-train and X-test*: Categorical input features underwent various encoding techniques, such as hashing IP and MAC addresses for efficient handling of string values and creating dummy variables through one-hot encoding for the Protocol column. These transformations enabled the capture of non-linear variable relationships, ultimately improving prediction accuracy.

3. Class Imbalance

Addressing class imbalance was another critical consideration. The datasets exhibited an uneven distribution of predicted values, with the normal class significantly outnumbering the suspicious class. To mitigate this imbalance, oversampling techniques were evaluated, with the Synthetic Minority Oversampling Technique (SMOTE) being selected as the preferred method. SMOTE's ability to generate unique synthetic samples for the minority class was advantageous, reducing overfitting risks and providing a more reliable representation of the minority class. Although other oversampling techniques were considered, SMOTE demonstrated the best balance between addressing class imbalance and preserving model sensitivity.

3.5   Feature Extraction and Selection

The study refrained from conducting separate feature selection and extraction procedures, as the network packet data captured using Wireshark was meticulously configured to provide a comprehensive set of essential attributes. These attributes encompassed key parameters, such as IP addresses, MAC addresses, ports, time to live, flags, ICMP code, and ICMP type. Wireshark's configuration offered direct access to these pertinent attributes, obviating the necessity for additional feature extraction processes. This approach allowed for the utilization of raw packet data, containing all requisite information for subsequent machine learning analyses.

3.6   Model Development

This subsection will outline the methodology adopted in the development of the advanced supervised machine learning model for smart grid cybersecurity. The process encompasses model selection, hyperparameter tuning, training, testing, evaluation, and the integration of the model with firewall rules for the prevention of malicious traffic. Each of these steps is vital in enhancing the security of smart grid control systems.

Model Selection. The model selection process involved careful consideration of several factors given the complexity of the datasets, including variables like IP addresses, MAC addresses, protocols, ports, and codes. Initially, four baseline models at their default hyperparameter settings were chosen: Random Forest (RF), Decision Tree (DT), Support Vector Machine (SVM), and Neural Network (NN) for their compatibility with non-linear data.

Performance evaluation revealed that DT and RF models tended to overfit the training data, impacting their performance on previously unseen data. In contrast, SVM and NN showed better generalization capabilities. Although NN had slightly lower accuracy on new data than SVM, it was chosen for its faster training process and lower memory requirements. To further enhance NN's performance, an ensemble approach with XGBoost was considered due to its regularization techniques and iterative decision tree refinement.

Hyperparameter Tuning. In the process of optimizing the proposed ensemble model after initial training and testing during the model selection phase, meticulous hyperparameter tuning was conducted [30]. For the XGBoost classifier, the study carried fine-tuned critical hyperparameters, including the learning rate and the number of estimators. The learning rate ranged from 0.01 to 0.3, and the number of estimators varied from 100 to 1000. This comprehensive exploration ensured a balance between model complexity and overfitting.

For the NN, hidden layer sizes of [50, 100, 50] were chosen to accommodate various network traffic patterns. An iteration limit of 1000 for convergence was set, the ReLu activation function for handling non-linearities was utilized, a small alpha value of 0.0001 for regularization was introduced, and a constant learning rate. Monitoring model performance during training was made possible with a validation fraction of 0.1.

The hyperparameters were optimized using an automatic grid search combined with cross-validation. This approach allowed us to explore a wide range of configurations, preventing overfitting and capturing intricate patterns in the complex network traffic data. The final hyperparameter settings are summarized in Table 2.

Table 2. Final hyperparameter setting values for optimal performance.

Parameter	Quantity
N estimators	100
Learning rate for XGBoost	0.1
Hidden layers	50, 100, 50
Iteration limit	1000
Activation function	ReLu
Validation function	0.1
Learning rate for NN	Constant
Alpha	0.0001

Voting Classifier to Ensemble Models. The model development culminated with the implementation of a Voting classifier. This ensemble model leveraged the unique strengths of individual models, combining their diverse decision-making approaches to enhance prediction accuracy. Using soft voting, the model considered the weighted average of predicted probabilities from XGBoost and NN, effectively mitigating biases from either model. While other ensemble techniques such as Bagging and Boosting were considered, the adaptability of the Voting classifier to network traffic patterns made it the preferred choice.

Model Detection Tasks - Proposed Model Training, Testing, and Evaluation. The ensemble model (NN and XGBoost) was meticulously trained on a 70% training set to capture intricate smart grid control systems network traffic patterns. Subsequently, the model underwent testing on a distinct 30% testing set to evaluate its generalization to real-world scenarios. The model was also tested on previously unseen data. Performance assessment included critical metrics like accuracy and detection rate [31]. Accuracy measured the model's capability in correctly classifying network traffic as either normal or suspicious. The detection rate measured the model's effectiveness in detecting suspicious traffic.

Model Prevention Tasks - Integration with Firewall Rules. Python programming was used to implement this whole code process, however, for demonstration purposes, algorithm 1 outlines a comprehensive procedure for seamlessly integrating the ensemble model with a firewall to enable the controlled passage of specific network traffic while preventing potentially harmful traffic. The initial phase involved loading a pre-trained ensemble model along with its dedicated preprocessor, thereby establishing a solid foundation for robust threat detection capabilities.

Subsequently, incoming unlabeled previously unseen data was subject to preprocessing and in-depth analysis, generating anomaly scores for each network packet. Those packets that exceeded a predefined threshold for anomaly scores were flagged as potential threats. A range of tests were performed, adjusting the threshold up and down, and evaluating the model's performance in blocking packets. An anomality threshold of 0.5 was set. A threshold of 0.5 performed well in identifying potential threats while maintaining an acceptable level of false positives. This triggered a dynamic process for crafting firewall rules tailored to the attributes of these packets, with the orchestration of this response facilitated by PowerShell and Python scripts.

Moreover, a logging system was used to record vital messages generated during packet processing. This logging mechanism ensured a comprehensive record for later review and analysis. Ultimately, this firewall algorithm served as a critical component of the ensemble model, taking proactive measures to safeguard the network by preventing the ingress of malicious traffic.

Algorithm 1. Firewall Integration Algorithm

Require: Trained model, automatic preprocessor
Ensure: Detection and prevention of suspicious network packets
    Initialize *Model*
    Initialize *Preprocessor*
    Initialize *UnseenData*
    $AnomalyScore \leftarrow Model.predict(Preprocessor.transform(UnseenData))$
    Set $Threshold \leftarrow 0.5$
    for each *Packet* in *UnseenData* **do**
        if $AnomalyScore[Packet] > Threshold$ **then**
            Mark *Packet* as a potential threat
            Extract *PacketAttributes* from *Packet*
            Create custom firewall rules based on *PacketAttributes* using PowerShell and Python scripts
            Execute the firewall rules with Subprocess module
        else
            Record information about the allowed *Packet* in the log file
        end if
    end for
    Initialize *LogFile*
    for each *Message* generated during processing **do**
        Append *Message* to *LogFile*
    end for
    Create a PowerShell script to add firewall rules based on *PacketAttributes*
    Execute the PowerShell script with Subprocess module

Evaluating Model Prevention Tasks. To evaluate how well the model was able to prevent and allow incoming traffic, the model was provided with unlabeled, unseen data for which the true labels were pre-known. Subsequently, the log file was assessed to examine closely the model's prevention process, specifically focusing on its accuracy in allowing and denying packets as well as focusing on packet information such as IP addresses, MAC addresses, ports, ICMP code, and ICMP type. These log files captured the model's decisions in real time as it processed incoming network traffic. The efficacy of the model's prevention results hinged on its ability to successfully detect traffic. In this context, prevention refers to the model's capacity to accurately identify and block malicious traffic while allowing legitimate traffic to pass unhindered.

4   Results and Analysis

In this section, the findings and results of the study will be presented, which aimed to fortify the control systems network against cyber threats by using an ensemble model.

4.1   Ensemble Model - NN and XGBoost

The results in Table 3 offer valuable insights into the performance of different models on testing sets, and unseen data and shed light on the efficacy of the proposed ensemble model consisting of NN and XGBoost.

1. Decision Trees (DT)
   DT exhibited a high accuracy of 99.41% on the testing set, indicating its ability to capture basic patterns in the data. However, this performance significantly dropped to 37.81% on unseen data, suggesting that DT struggled to generalize to new and complex patterns. This indicates overfitting, where the model learned the training data too well but failed to adapt to unseen instances effectively.

2. Random Forest (RF)
   RF, as an ensemble model, achieved a relatively high accuracy of 99.58% on the testing set, even outperforming DT. However, its performance on unseen data remained subpar at 38.72%, echoing the overfitting issue observed with DT. RF's ensemble nature did not mitigate this overfitting problem, resulting in a significant performance gap.

3. Support Vector Machines (SVM)
   SVM exhibited a high accuracy of 97.77% on the testing set, indicating its
   generalization capability. However, it performed remarkably well on previously unseen data with an accuracy of 89.67%, suggesting that it excelled in uncovering concealed subtleties within the network traffic data. Nonetheless, its performance on testing data was slightly lower compared to DT and RF.

4. NN and XGBoost Ensemble
   The ensemble model NN and XGBoost displayed remarkable performance, achieving a perfect accuracy of 97.89% on the testing set. More importantly,

this performance translated well to unseen data, with an accuracy of 93.19%. This suggests that the NN, with its ability to unravel intricate patterns, complemented XGBoost's gradient-boosting prowess effectively. The ensemble's robust performance on both testing and unseen data showcases its potential as a reliable solution for smart grid control systems network security.

Table 3. Model selection - performance results

Model	Accuracy on Testing set (%)	Accuracy on Unseen data (%)
DT	99.41	37.81
RF	99.58	38.72
SVM	97.77	89.67
NN and XGBoost	97.89	93.19

4.2   Hyperparameter Tuning Effect on Model Detection Tasks

Hyperparameter tuning played a pivotal role in optimizing the ensemble model, significantly enhancing its performance. The process focused on fine-tuning key parameters for both the XGBoost classifier and the NN, resulting in substantial improvements in model accuracy and detection rate. (see Fig. 4) illustrates the changes in accuracy and detection rate before and after hyperparameter tuning.

Initial Model Performance: Before hyperparameter tuning, the model already demonstrated a high level of performance, with an accuracy of 97.89% and a detection rate of 93.19% on the testing set. While these results were commendable, they left room for further improvement.

Hyperparameter Tuning Results: After meticulous hyperparameter tuning, the model's performance saw a remarkable enhancement. The accuracy of the model increased to 100%, while the detection rate improved to 99.19%. These improvements highlight the effectiveness of hyperparameter tuning in capturing intricate patterns within complex network traffic data.

Quantitative Improvements: Hyperparameter tuning resulted in a 2.28% increase in accuracy and a 6.44% improvement in the detection rate. These quantitative results underscore the significant impact of parameter fine-tuning in creating a highly accurate and effective model for detection tasks.

Context and Challenges: It is essential to recognize that, despite the initial high-performance levels, the intricate nature of network traffic data presented specific challenges, particularly in the realm of detecting previously unseen data. Initial results indicated an accuracy of 93.19% for identifying previously unobserved network traffic patterns. These challenges stemmed from the dynamic and evolving nature of network traffic, which demanded a more refined approach. The process of hyperparameter tuning was instrumental in addressing these unique challenges by optimizing the model's parameters for optimal performance. As a result, the accuracy in identifying previously unseen data surged to 99.19%, showcasing the profound impact of hyperparameter tuning in enhancing the model's ability to cope with complex and ever-changing network traffic data patterns.

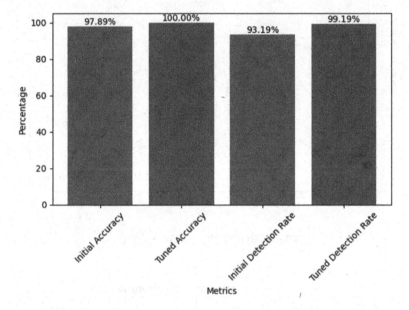

Fig. 4. Performance Before and After Hyperparameter Tuning on Testing Set

4.3 Receiver Operating Characteristic Area Under the Curve on Testing and Unseen Data

The ROC AUC [32] value of 1 on the testing set (See Fig. 5) suggested that the model achieved perfect discrimination between classes. This means that the model separated the classes, resulting in no false positives and no false negatives on the testing data. Therefore, indicated a well-optimized model that accurately captured patterns in the training data.

The ROC AUC value is 0.99 on unseen data See Fig. 6, which is very close to 1. This indicated that your model's performance remained excellent even when dealing with new, previously unseen data. High ROC AUC value on unseen data was the evidence that the model generalizes well and is robust in real-world scenarios. The model's ability to maintain such high performance on unseen data not only demonstrated its robustness but also highlighted its practical applicability in network traffic analysis. This quality was essential for addressing the dynamic nature of network traffic and ensured accurate and reliable results in various operational settings.

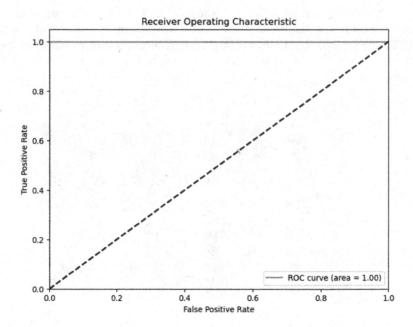

Fig. 5. ROC AUC on Testing Set

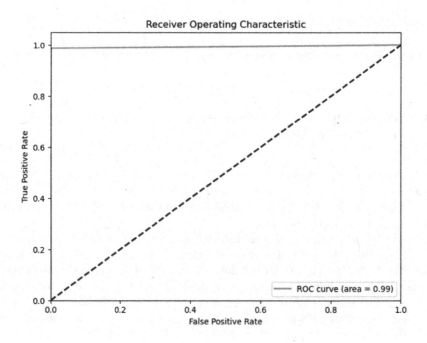

Fig. 6. ROC AUC on Unseen Data

4.4   Proposed Model Versus Existing Models

To further evaluate the performance of the proposed ensemble model for smart grid control network traffic analysis, the results of the proposed model were compared with those of existing ensemble models. Table 4) presents a summary of the comparison based on the accuracy and detection rates achieved by each model.

Table 4. Model performance against existing models

Algorithms Approach	Accuracy (%)	Detection (%)
Supervised and Unsupervised model [15]	–	73.0
Random Forest [16]	99.4	99.2
Extremely Randomised Regression [17]	97.0	–
Probabilistic Bayesian Regression [18]	99.0	99.0
Firewall Model with CART [19]	96. 0	96.0
Proposed Ensemble Model (NN and XGBoost), its average score on testing and unseen data	99.6	99.48

It is evident from the comparison that the proposed ensemble model achieved competitive accuracy and detection rates, and outperformed some existing models in both aspects. Additionally, the model showcased robustness by achieving high detection rates without relying solely on a high accuracy rate.

4.5   Evaluation of Prevention Tasks

The model's prevention function was elucidated through an in-depth analysis of the comprehensive log file (see Fig. 7). This log meticulously captured and documented both authorized and denied network packets, encompassing critical parameters like source and destination IP addresses, ports, ICMP types, and codes. The model firewall capability showed remarkable strength becomes evident in its precision in detecting capabilities, effectively distinguishing between legitimate and potentially malicious communication. (See Fig 5) instances of permitted communication, exemplified by interactions between 10.10.10.3 (Control-PC 1) and 10.10.10.6 (SCADA Server), thus affirming the successful filtration of legitimate network traffic. Conversely, the firewall exhibited its efficacy by correctly identifying and rejecting DHCP starvation traffic. These results accentuate the model's outstanding performance in safeguarding the network against a wide spectrum of cyber threats.

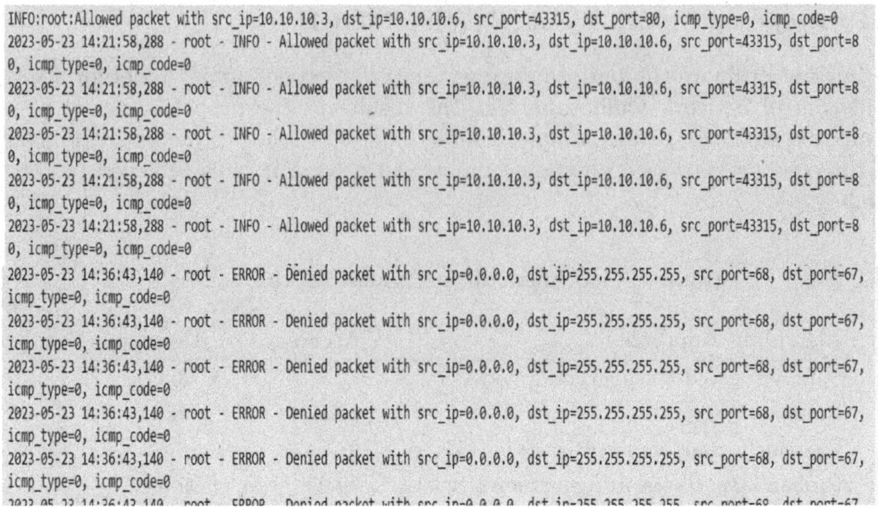

```
INFO:root:Allowed packet with src_ip=10.10.10.3, dst_ip=10.10.10.6, src_port=43315, dst_port=80, icmp_type=0, icmp_code=0
2023-05-23 14:21:58,288 - root - INFO - Allowed packet with src_ip=10.10.10.3, dst_ip=10.10.10.6, src_port=43315, dst_port=8
0, icmp_type=0, icmp_code=0
2023-05-23 14:21:58,288 - root - INFO - Allowed packet with src_ip=10.10.10.3, dst_ip=10.10.10.6, src_port=43315, dst_port=8
0, icmp_type=0, icmp_code=0
2023-05-23 14:21:58,288 - root - INFO - Allowed packet with src_ip=10.10.10.3, dst_ip=10.10.10.6, src_port=43315, dst_port=8
0, icmp_type=0, icmp_code=0
2023-05-23 14:21:58,288 - root - INFO - Allowed packet with src_ip=10.10.10.3, dst_ip=10.10.10.6, src_port=43315, dst_port=8
0, icmp_type=0, icmp_code=0
2023-05-23 14:21:58,288 - root - INFO - Allowed packet with src_ip=10.10.10.3, dst_ip=10.10.10.6, src_port=43315, dst_port=8
0, icmp_type=0, icmp_code=0
2023-05-23 14:36:43,140 - root - ERROR - Denied packet with src_ip=0.0.0.0, dst_ip=255.255.255.255, src_port=68, dst_port=67,
icmp_type=0, icmp_code=0
2023-05-23 14:36:43,140 - root - ERROR - Denied packet with src_ip=0.0.0.0, dst_ip=255.255.255.255, src_port=68, dst_port=67,
icmp_type=0, icmp_code=0
2023-05-23 14:36:43,140 - root - ERROR - Denied packet with src_ip=0.0.0.0, dst_ip=255.255.255.255, src_port=68, dst_port=67,
icmp_type=0, icmp_code=0
2023-05-23 14:36:43,140 - root - ERROR - Denied packet with src_ip=0.0.0.0, dst_ip=255.255.255.255, src_port=68, dst_port=67,
icmp_type=0, icmp_code=0
2023-05-23 14:36:43,140 - root - ERROR - Denied packet with src_ip=0.0.0.0, dst_ip=255.255.255.255, src_port=68, dst_port=67,
icmp_type=0, icmp_code=0
2022 05 23 14:36:43 140   root   ERROR   Denied packet with src_ip=0.0.0.0  dst_ip=255 255 255 255  src_port=68  dst_port=67
```

Fig. 7. Log file analysis - Model prevention tasks

5   Discussion and Implications

The ensemble model, combining the strengths of NN and XGBoost, addresses challenges in smart grid control systems network traffic analysis effectively. NN [33], with deep learning capabilities, excels at unraveling complex patterns in network traffic data. Its multiple hidden layers and nonlinear activation functions enhance its ability to detect anomalies. On the other hand, XGBoost [34], a gradient boosting algorithm, demonstrates exceptional prowess in classification tasks by iteratively training decision trees to improve predictive power.

The Voting classifier, amalgamating NN and XGBoost decision-making, enhances prediction robustness, mitigating biases. The model achieves a remarkable 99.19% accuracy on unseen data.

Hyperparameter tuning involves adjustments to various model parameters, crucial for observed enhancements. Each hyperparameter's impact on model performance was carefully considered.

As a firewall, the model represents a significant advancement in smart grid control systems network security. Its dynamic analysis and packet filtering, driven by predictive capabilities, offer the potential for safeguarding critical infrastructures. The firewall's ability to differentiate between normal and anomalous behaviors is advantageous in smart grid control networks.

Strategically positioned at the network perimeter, the model acts as the first line of defense against incoming threats, scrutinizing and promptly blocking suspicious activities. This placement safeguards internal systems, extending protection to control centers, substations, and field devices, creating a comprehensive defense system.

The research findings can influence policy development in the smart grid control systems field. Policymakers and industry stakeholders should consider adopting predictive machine learning models, like the proposed ensemble model, to enhance smart grid network security, contributing to a safer and more reliable energy supply.

6   Conclusion

In conclusion, the ensemble model, which combines the strengths of the NN and XGBoost, has proven to be a robust and reliable solution for addressing the intricate challenges of smart grid control systems network traffic analysis. The deep learning capabilities of NN enable it to uncover complex patterns within network traffic data, while XGBoost's gradient-boosting algorithm excels in classification tasks. The Voting classifier, which combines the decision-making strategies of both NN and XGBoost, enhances the model's resilience and prediction accuracy of 99.19% and detection rate of 99.48% on average.

When evaluated as a firewall, the model offers a significant advancement in network security. Its dynamic analysis and packet filtering based on predictive insights hold great potential for safeguarding critical infrastructures. The model's ability to distinguish between normal and anomalous behaviors is a key advantage in the complex landscape of smart grid control networks.

The model's effectiveness at the network perimeter, as the first line of defense against incoming threats, ensures the protection of internal systems. This approach extends its protective measures to control centers, substations, and field devices, creating a comprehensive defense system for critical infrastructures.

7   Recommendations and Future Work

To further enhance the robustness and reliability of the ensemble model and its application as a firewall for smart grid control systems, several recommendations are suggested. Firstly, continuous monitoring and updates of the model and firewall rules should be implemented to keep pace with evolving cyber threats and maintain their effectiveness. This includes staying informed about the latest attack techniques and vulnerabilities and adapting the model accordingly.

Additionally, the integration of a real-time response system is recommended to enable immediate counteraction against detected threats. Automating responses based on anomalies identified by the model can bolster network security and reduce response time.

Scalability is another crucial consideration. As smart grid networks expand, the model should be designed to handle larger and more complex networks without sacrificing performance. Ensuring its scalability will be vital for its long-term utility.

Developing a user-friendly interface for network administrators to interact with the firewall and monitor its activities is also advisable. This would make the model and firewall more accessible and practical for those responsible for managing smart grid networks.

Conducting extensive testing in real-world environments is essential to validate the model's effectiveness and reliability. Collaborating with power utilities and industries that operate smart grids can provide valuable insights and data for further model improvement.

In conclusion, implementing these recommendations will contribute to the continued success and applicability of the ensemble model and firewall in securing smart grid control systems and critical infrastructure against cyber threats.

Acknowledgments. A special thanks to the Center for AI Research (CAIR), School of Data Science and Computational Thinking, and Department of Information Science, Stellenbosch University for their support throughout this research.

References

1. Borenius, S., Gopalakrishnan, P., Bertling, L., Kantola, R.: Expert-guided security risk assessment of evolving power grids. Energies **15**(9), 3237 (2022)
2. Albright, D., Brannan, P., Walrond, C.: Stuxnet malware and Natanz: update of ISIS December 22, 2010 report. Inst. Sci. Int. Secur. **15**, 739883–3 (2011)
3. Liang, G., Weller, S.R., Zhao, J., Luo, F., Dong, Z.Y.: The 2015 Ukraine Blackout: implications for false data injection attacks. IEEE Trans. Power Syst. **32**(4), 3317–3318 (2016)
4. Markos, Y.: Cyber Security Challenges that Affect Ethiopia's National Security. SSRN 4190146 (2022)
5. Panettieri, J.: Ransomware Attack Rocks City Power, Johannesburg, South Africa. MSSP Alert (2019). www.msspalert.com/cybersecurity-breaches-and-attacks/ransomware/city-power-johannesburg-south-africa. Assessed 24 Apr 2023
6. Abrahamsen, F.E., Ai, Y., Cheffena, M.: Communication technologies for smart grid: a comprehensive survey. Sensors **21**(23), 8087 (2021)
7. Vahidi, S., Ghafouri, M., Au, M., Kassouf, M., Mohammadi, A., Debbabi, M.: Security of wide-area monitoring, protection, and control. (WAMPAC) Systems of the Smart Grid, A Survey on Challenges and Opportunities. IEEE Communications Surveys and Tutorials (2023)
8. Bharadiya, J.: Machine learning in cybersecurity: techniques and challenges. Eur. J. Technol. **7**(2), 1–14 (2023)
9. Koay, A.M., Ko, R.K.L., Hettema, H., Radke, K.: Machine learning in industrial control system (ICS) security: current landscape, opportunities and challenges. J. Intell. Inf. Syst. **60**(2), 377–405 (2023)
10. Farrar, N.O., Ali, M.H., Dasgupta, D.: Artificial intelligence and machine learning in grid connected wind turbine control systems: a comprehensive review. Energies **16**(3), 1530 (2023)
11. Surucu, O., Gadsden, S.A., Yawney, J.: Condition monitoring using machine learning: a review of theory, applications, and recent advances. Expert Syst. Appl. **221**, 119738 (2023)

12. Dehghani, M., et al.: Blockchain-based securing of data exchange in a power transmission system considering congestion management and social welfare. Sustainability **13**(1), 90 (2020)
13. Saha, S.S., Gorog, C., Moser, A., Scaglione, A., Johnson, N.G.: Integrating hardware security into a blockchain-based transactive energy platform. In: 2020 52nd North American Power Symposium (NAPS), pp. 1–6, April 2021
14. Zhang, H., Wang, J., Ding, Y.: Blockchain-based decentralized and secure keyless signature scheme for Smart Grid. Energy **180**, 955–967 (2019)
15. Ashrafuzzaman, M., Das, S., Chakhchoukh, Y., Shiva, S., Sheldon, F.T.: Detecting stealthy false data injection attacks in the smart grid using ensemble-based machine learning. Comput. Secur. **97**, 101994 (2020)
16. Zidi, S., Mihoub, A., Qaisar, S.M., Krichen, M., Al-Haija, Q.A.: Theft detection dataset for benchmarking and machine learning based classification in a smart grid environment. J. King Saud Univ. Comput. Inf. Sci. **35**(1), 13–25 (2023)
17. Acosta, M.R.C., Ahmed, S., Garcia, C.E., Koo, I.: Extremely randomized trees-based scheme for stealthy cyber-attack detection in smart grid networks. IEEE Access **8**, 19921–19933 (2020)
18. Soltan, S., Mittal, P., Poor, H.V.: Line failure detection after a cyber-physical attack on the grid using Bayesian regression. IEEE Trans. Power Syst. **34**(5), 3758–3768 (2019)
19. Haghighi, M.S., Farivar, F., Jolfaei, A.: A machine learning-based approach to build zero false-positive IPSs for industrial IoT and CPS with a case study on power grids security. IEEE Trans. Ind. Appl. **60**, 920–928 (2020)
20. Choi, B.: Introduction to VMware workstation. In: Introduction to Python Network Automation: The First Journey, pp. 139–168. Apress, Berkeley, CA (2021)
21. Cesar, P., Pinter, R.: Some ethical hacking possibilities in Kali Linux environment. J. Appl. Techn. Educ. Sci. **9**(4), 129–149 (2019)
22. Soepeno, R.A.A.P.: Wireshark: An Effective Tool for Network Analysis (2023)
23. Raschka, S.: Python Machine Learning: Machine Learning and Deep Learning with Python, scikit-learn, and TensorFlow (2018)
24. Banda, T.V.: Towards a Supervised Machine Learning Algorithm for Cyberattacks Detection and Prevention in a Smart Grid Cybersecurity System. Stellenbosch University (2023)
25. Stouffer, K., Falco, J., Scarfone, K.: Guide to industrial control systems (ICS) security. NIST Spec. Publ. **800**(82), 16–16 (2011)
26. Allen, L., Heriyanto, T., Ali, S.: Kali Linux-Assuring Security by Penetration Testing. Packt Publishing Ltd. (2014)
27. Denis, M., Zena, C., Hayajneh, T.: Penetration testing: concepts, attack methods, and defense strategies. In: 2016 IEEE Long Island Systems, Applications and Technology Conference (LISAT), pp. 1–6. IEEE, April 2016
28. Orebaugh, A., Pinkard, B.: Nmap in the Enterprise: Your Guide to Network Scanning. Elsevier (2011)
29. Mishra, P., Biancolillo, A., Roger, J.M., Marini, F., Rutledge, D.N.: New data preprocessing trends based on ensemble of multiple preprocessing techniques. TrAC, Trends Anal. Chem. **132**, 116045 (2020)
30. Yang, L., Shami, A.: On hyperparameter optimization of machine learning algorithms: theory and practice. Neurocomputing **415**, 295–316 (2020)
31. Vujović, Z.: Classification model evaluation metrics. Int. J. Adv. Comput. Sci. Appl. **12**(6), 599–606 (2021)
32. Muschelli III, J.: ROC and AUC with a binary predictor: a potentially misleading metric. J. Classif. **37**(3), 696–708 (2020)

33. Abiodun, O.I., Jantan, A., Omolara, A.E., Dada, K.V., Mohamed, N.A., Arshad, H.: State-of-the-art in artificial neural network applications: a survey. Heliyon **4**(11), 1–41 (2018)
34. Ramraj, S., Uzir, N., Sunil, R., Banerjee, S.: Experimenting XGBoost algorithm for prediction and classification of different datasets. Int. J. Control Theory App. **9**(40), 651–662 (2016)

Classification of DGA-Based Malware Using Deep Hybrid Learning

Bereket Hailu Biru[1,2]([✉])[iD] and Solomon Zemene Melese[2,3][iD]

[1] Ethiopian Artificial Intelligence Institute, 40782 Addis Ababa, Ethiopia
bek.hailu@gmail.com
[2] Department of Electrical and Computer Engineering, Addis Ababa Science
and Technology University, 16417 Addis Ababa, Ethiopia
solwub16@gmail.com
[3] Center of Excellence for High-Performance Computing and Big Data Analysis,
AASTU, Addis Ababa, Ethiopia

Abstract. Attackers use a domain generation algorithm (DGA) to generate a large number of random domain names that act as rendezvous points. These domain names create a command-and-control channel between the attackers and their malware, enabling the hackers to send any command to the malware. Detecting such domain names is challenging as they mimic the pattern of normal domain names, making it difficult to attribute specific malware-generating domains. Moreover, detecting DGA-based domain names faces the issue of poor performance in detecting zero-day malware. Our study uses a hybrid deep learning model with machine learning algorithms that automatically learn features from given domain name data. We used CNN and LSTM deep learning algorithms as feature extractors and seven machine learning algorithms, namely Logistic Regression (LR), Naive Bayes (NB), Decision Tree (DT), Random Forest (RF), Extra Trees (ET), AdaBoost (AB), and Extreme Gradient Boosting (XGB), for the final classification experiments to solve both binary and multiclass problems. Our findings indicate that the hybrid model of LSTM with the random forest algorithm scored 99% detection accuracy and 99.8% AUC score in binary experiments and detection accuracy of 99% and 99.65% AUC score in grouped multiclass classification experiments. Grouping domain families based on their character distribution gives better results than trying to detect families individually.

Keywords: DNS(Domain Name System) · Malware · DGA(Domain Generation Algorithm) · DHL(Deep Hybrid Learning) · C&C server

1 Introduction

The frequency of cyber invaders and hijackers is increasing daily, which poses a high risk and can result in significant destruction in the digital world. Even if a network has the highest level of security, it becomes vulnerable to attack

T. G. Debelee et al. (Eds.): PanAfriConAI 2023, CCIS 2069, pp. 129–150, 2024.
https://doi.org/10.1007/978-3-031-57639-3_6

once connected to the internet, as the internetwork is not designed to be secure. It is crucial to address both network and internetwork security, since when it comes to internetwork security breaches, DNS is often the key target [1]. DNS acts as the phone book for digital assets, holding the domain name to an IP address map. It defines who is online and reveals company identities. If someone gains control of a company's DNS, they can compromise anything connected to it [1]. The Domain Security Report released by CSC on June 23, 2020, revealed that 83 out of 2000 global organizations are at a higher risk of domain name hijacking [2]. DNS abuse can redirect employees to malware-infected sites, grant unauthorized access to company servers, and result in the theft of sensitive data [3]. DNS attacks are challenging to prevent as attackers use various methods and techniques to carry out their malicious activities. Traditional approaches to preventing DNS abuse have failed due to the use of domain generation algorithm tools by attackers, making it difficult to reverse engineer the attacks. The Domain Generation Algorithm (DGA) is a code that serves two primary purposes. Firstly, it generates a large number of domain names randomly during an attack to bypass the defense mechanism that relies on blocking a fixed domain. This technique enables attackers to switch between domains and makes it challenging to block or remove them. Secondly, DGA creates command and control channels between malware and attackers, allowing them to issue commands from the hacker's server (C2 server). Cybercriminals utilize DGA as a tool to generate a massive list of new, random domains to evade security countermeasures. Attackers create and hide C2 servers to avoid detection by standard cybersecurity methods by generating a large number of domain names as needed. DNS security faces a major challenge in detecting DGA. Malware uses different seeds to generate domain names that imitate the pattern of normal domain names [4]. They do this by concatenating pseudo-randomly chosen English dictionary words. The intention is to conceal their activities and avoid detection. DNS was originally designed for name resolution, not data transfer [5,6]. However, cybercriminals have exploited it to exfiltrate sensitive information without the knowledge of users. Hackers are constantly seeking new ways to exploit the DNS protocol and infrastructure. According to a report by therecord.media, the number of C&C servers has increased by 30% in 2022. In December 2021, there were over 17,000 units, compared to 13,629 in 2021 [7]. Since generated DGA domains are only used for a short time, it becomes challenging to find links between samples. One way to tackle this issue is by using AI-based models that can learn from new data and improve their accuracy in identifying DGA-based malware. These models can analyze patterns and behaviors associated with DGAs, improve detection intelligence and security solutions, and provide insight for security vendors. However, several challenges hinder the detection mechanisms, such as the inability to attribute specific malware-generating domains, poor performance in detecting zero-day malware, poorly organized or reviewed datasets [3,8], improper selection of learning and classification algorithms, and insufficient context and features used to express the domain name [3,9,10].

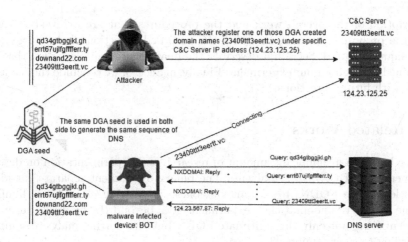

Fig. 1. How hackers use DGA to generate fake DNS and create C&C channels: Both the client and source sides understand the seed, resulting in identical sequences being generated without the need for communication. This knowledge allows a hacker to anticipate the domain name that the malware will use. The attacker then registers a domain from the sequence, creating a channel or rendezvous point between the malware and the hacker [11].

As per Netlab 360, there are over 50 DGA malware families based on domain generation algorithms (DGAs). Some DGAs have strange combinations of letters, like gegjiimqmlgtdmk.tf or jxbdxeyxttdmcjagi.me, while others are more challenging to detect, such as huoseavas.name or agtisaib.info, as they are designed to look like human language. To avoid being blocked or sinkholed, DGAs use input seeds to randomize their output and make it unpredictable. These seeds can be things like the current date and time, values embedded into a sample group by a campaign, or strings obtained from malware authors or public servers. Domain name families created using English words (like the Matsnu and Suppobox families) can be harder to detect, depending on their lexical features [8].

This study's goal is to enhance the DGA-based malware classification system. The contributions to this paperwork will be:

1. Deep hybrid model for DGA-based malware classification using both machine learning and deep learning classifiers.
2. Test the trained model more extensively on domain names generated by new and previously unseen (i.e., untrained) malware.
3. The proposed approach is evaluated using different evaluation metrics, and the results are compared with those of other existing methods.

The paper is organized into several sections. The first section, Sect. 1, contains the introduction, which explains the basic information, objective, and contribution of the thesis. In Sect. 2, the related works based on the proposed work of DGA attack classification mechanisms are discussed. Section 3 contains a description of the overall work and the procedures that are followed to carry out the proposed research, including the proposed detection methodology and techniques adopted. Section 4 describes the final results and discussion that we obtained throughout our experiment. Finally, in Sect. 5, we conclude the research work with future directions.

2   Related Works

There has been a significant amount of research and experimentation on detecting benign or DGA-based malware. The binary experiment separates or identifies legitimate FQDNs from malicious ones, considering all malware families as one category. The multiclass experiment goes beyond the binary experiment by identifying not only the legitimate FQDN but also sorting malware samples according to their families [3].

Recently published methods, such as FANCI (a random forest based on human-engineered features) and LSTM.MI (a deep learning approach), was tested against a unique DGA called CharBot, which was presented in cite8756038. The results showed poor performance, revealing a dangerous weakness of modern DGA classifiers. They are vulnerable to extremely simple attacks that do not use sophisticated machine-learning techniques [12].

The authors in [8] combined a novel recurrent neural network architecture with domain registration side information (WHOIS) to detect DGA domains. Their experiment detected DGA families with high smash-word scores but with low accuracy and was unable to detect DGA families that do not look like natural domain names. Since DGA-based malware uses a variety of seeds, developing a model capable of detecting malicious domains altogether is critical, and every one of the models tested here fails to do so [13].

Paper [14] proposed a parallel hybrid architecture named Bilbo, composed of an LSTM, a CNN, and an ANN, for dictionary DGA detection, which is harder to detect due to their natural language characteristics. Their architecture was trained and tested on Alexa and three selected dictionary DGA datasets. The classification result revealed that their model was able to detect dictionary-based DGA in multiclass classification. However, their BILBO model scored an FPR of 4.54% in the binary experiment.

In [15], they used a Long Short-Term Memory (LSTM) with an attention mechanism method for DGA domain name classification. They used the character sequence of the domain name as a feature. The experimental results show that combining the attention mechanism with the LSTM can effectively classify DGA domain names, but no significant improvement is achieved.

In [4], the authors presented an effective ATT-CNN-BiLSTM model that integrated an attention mechanism and deep neural network to detect and classify domain names. They achieved better performance on arithmetic-based, part-wordlist-based, and wordlist-based DGAs like 'matsnu' and 'suppobox' families. However, malware like 'Cryptolocker', 'gameover', and 'locky' are not correctly classified.

In [16], K. P. proposed a deep-learning approach to DGA classification to detect DGAs that generate malicious domains randomly. With the help of additional feature extraction and knowledge-based extraction in the deep learning architecture (CNN and RNN), they conducted the binary experiment and achieved 95.9% accuracy for DGA classification.

In related work done on DGA-based malware detection, a different approach to DGA-based malware classification was studied. Most experiments were unsuccessful in the identification of some malware, especially in multiclass experiments. Additionally, there is poor performance in the detection of zero-day (previously unseen) malware. Deep learning algorithms need a large amount of training data to make a correct prediction through their data processing, and class imbalances in a collected data set lead a model to a biased classification. These are the reasons behind the gap observed in the reviewed work, in addition to the inability of manually engineered features to represent the input domain data needed to implement machine learning algorithms.

3 Methodology

3.1 Proposed Methods

The performance and accuracy of classical machine learning models are often limited by the dataset and feature engineering done on [17]. On the other hand, using fully connected neural network layers for the final classification or clustering layer in a deep learning model may result in overfitting with fewer data points or require excessive computational resources. To overcome these limitations, the proposed research will use a deep hybrid approach (ML and DL) that combines the benefits of both approaches. By doing so, the prediction accuracy can be increased while the computational complexity can be reduced [15] (Fig. 2).

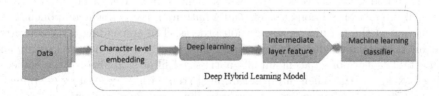

Fig. 2. Deep hybrid model.

3.1.1 Deep Hybrid Learning (DHL).

Deep hybrid learning is the result of fusion networks, which can be obtained by combining deep learning and machine learning techniques [18]. Thus, we can take advantage of the benefits of both DL and ML, reduce the drawbacks of both techniques and provide more accurate and less computationally expensive solutions by using deep hybrid learning (DHL). CNN and LSTM deep learning algorithms were selected as feature extraction algorithms for our classification task. CNN can handle noisy, misspelled, and out-of-vocabulary words, whereas LSTM can learn and memorize long-term dependencies, i.e., it can retain past information over time. Due to this, LSTM is effective in time series prediction. The seven machine learning algorithms-Decision Tree, Logistic Regression, Naive Bayes, Random Forest, AdaBoost, XGBoost, and Extra Tree-were used for the final classification tasks. The features extracted from an intermediate layer of CNN and LSTM were used as input for those machine learning algorithms (Fig. 3).

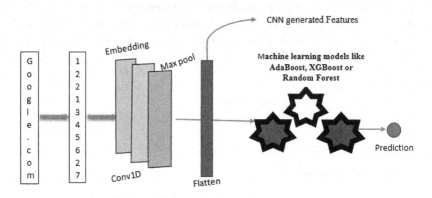

Fig. 3. Deep Hybrid model with CNN.

3.1.2 1D CNN(Convolutional Neural Network).   CNN is a multi-layered artificial neural network that can detect complex features in data. Text-based CNN is applied to extract local information from the characters of each input domain name [12]. 1D-CNN is a commonly used architecture of CNN for natural language processing or for analysis of the time sequence of sensor data. The proposed CNN network consists of an embedding layer, a 1D-sized convolutional filter, a Max Pooling layer, and a flattening layer. Each component has a function in the feature representation process. The embedding layer converts the input character sequence vector into a matrix of N x M. Then, to capture an important feature of adjacent inputs, a 1D-sized filter is applied to the embedding N x M matrix. Then, a down-sampling step with Max Pooling is applied. This maximum pooling operation performs the dimensional reduction by selecting the maximum value of the input data to create the new feature map. Finally, a flattening operation is performed for all feature maps to extract a single and a new feature vector (Fig. 4).

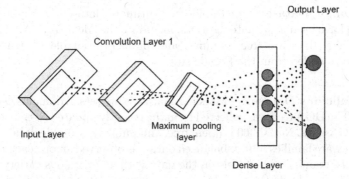

Fig. 4. 1D Convolutional Neural Networks model.

3.1.3 LSTM (Long Short-Term Memory).

LSTM is a kind of recurrent neural network (RNN) that can learn and memorize long-term dependencies. The default behavior of the LSTM network is to recall past information for long periods. They retain information over time. LSTM models can also be applied for extracting character-level features. We apply the LSTM to the sequence of character embeddings for each input domain. Character-Level-LSTM Implemented a character-level RNN that learns from some text one character at a time. The LSTM model effectively captures the global features of each word token. Then the flattening layer operation is applied to extract a new feature vector.

3.1.4 Machine Learning Algorithms.

Machine learning algorithms require guidance on how to make accurate predictions by processing more information, which can be achieved through feature extraction. These algorithms can make predictions using small amounts of data and can operate on low-end machines without requiring a significant amount of computational power and time. However, it is essential that users accurately identify and create features.

In our case, we utilized a deep learning algorithm as a feature extractor and machine learning algorithms for the final classification task. For the final classification experiments, we employed seven machine learning algorithms, namely Logistic Regression (LR), Naive Bayes (NB), Decision Tree (DT), Random Forest (RF), Extra Trees (ET), AdaBoost (AB), and Extreme Gradient Boosting (XGB), to solve both binary and multiclass problems.

3.2 Dataset

In this research, "Malicious" (DGA), "Benign" (Non-DGA), and unseen malware families that were not included in the training and testing sets were collected from different sources separately to conduct our experiment.

3.2.1 Benign Dataset: Samples in legitimate datasets are real domain names. We have used legitimate datasets, including the Alexa Top 1M legitimate domain. These data are free, popular, and publicly available benign domain datasets downloaded from the Kaggle site.

3.2.2 Malicious Dataset: Malicious datasets are sets of domains generated by DGAs. The DGA feed was collected from the recently updated Netlab360 DGA Feeds as DGA data. Netlab360 is open, free, and publicly available data that contains 56 malware families, but unbalanced data was observed among each malware family. Thus, training our models on the data with an enormous variety of malware families would help the models learn a variety of malware features since they are generated from a variety of seeds. This has contributed to the performance of the model when tested on previously unseen, novel malware families.

3.2.3 Unseen Dataset: These datasets were collected from DomCop Legit and DGArchive for zero-day malware detection experiments and to test the model on new families. This data contains 17 DGA families and a legit domain with a total of 1,339,087 records (Table 1).

Table 1. The collected Data size.

Dataset	Data size
Netlab360	1,488,595
Alexa Top 1M domains	613,032
Unseen Data	1,339,087

3.3 Data Preprocessing

Data preprocessing is a data mining technique that deals with data preparation to improve the quality of the data and make it suitable for the designed models. It is an important step to enhance data efficiency. The accuracy of the machine learning model is directly correlated with the quality of training data, i.e., the quality of data always affects the final prediction of models in experimental results. Techniques such as data cleaning, labeling, encoding, feature extraction, and embedding are used to enhance data quality and efficiency. The ultimate goal is to obtain accurate and reliable results in any experiment.

3.3.1 Data Labeling: Data labeling is the process of adding tags or labels to the collected row data. Our data, which were collected from different sources, were raw, unlabeled, and unfeatured datasets. Since the classification problem has been identified, it is appropriate to implement supervised machine learning algorithms. Supervised machine learning algorithms require labeled data, and they learn under the supervision of labeled data. Thus, it needs to label the domain data samples as benign and malicious for a binary experiment and also label each domain based on its family for a multiclass experiment. Therefore, our collected benign and malicious data were labeled and concatenated with each other for both binary and multiclass classification problems separately.

3.3.2   Data Cleaning: Data cleaning is the process of identifying and removing duplicates and unwanted or irrelevant data samples from a given data sample. Row data are highly vulnerable to missing, duplication, outliers, and inconsistencies and may have incomplete records, duplicate values, and noise. Unclean data can result from human error, scraping data, or combining datasets from multiple sources. This issue can inevitably skew the data and confuse the final result. Thus, data cleaning is an important step to recognize and deal with null, duplicate, corrupted, unwanted characters, and wrongly formatted data frame values. This kind of data issue can be handled by using different techniques. When it comes to the missing value, removing the observation that has the missing value or inputting the missing data can be used as a solution. When it comes to duplication, removing it is the solution. In our labeled data, we have tried to identify the null and duplicate values and handle the issues by removing duplicate values. The total data before cleaning was 2,121,627; after the cleaning process, the entire data left was 2,076,638.

3.3.3   Encoding Technique   Machine learning algorithms operate on numerical data only. They require the input and output values to be numeric to perform a mathematical operation on the data. Encoding is a method to deal with categorical data in a dataset. It is very important to convert categorical data (primarily in string form) into numerical values since many machine learning algorithms accept only numerical values. There are different encoding techniques in machine learning based on the data type. When it comes to ordinal data, the data that has an explicit ordering, label encoder, and label binarizer is a suitable technique that we applied to our ordinal data to encode labels.

1. Label encoder: is one of the encoding techniques from the Python sci-kit-learn library that is used for encoding the target, i.e., the output label. Simply put, it assigns an integer value to every possible value of a categorical variable. This technique can be applied to both binary and multi-class problems. In this encoding technique, the categorical data is assigned a value from 1 to N, where N is the number of different categories present in the data. We used this encoding technique for both binary and multiclass experiments with ensemble learning algorithms.
2. Label binarization is an easy process to convert multi-class labels to binary labels (belong to or do not belong to the class). It binarizes labels in one-vs.-all fashion. It becomes possible to compute this transformation for a fixed set of class labels known ahead of time. We implemented this method for our proposed deep hybrid models in a multiclass classification experiment to encode the target labels.

3.3.4 Class Balancing: refers to the process of balancing the classes of unbalanced samples in a dataset. This issue arises when there is an unequal representation of classes within the data, which can occur in both binary and multiclass classification scenarios [19]. It is crucial to avoid class imbalances before applying machine learning algorithms to ensure unbiased classification. To address this concern, we utilized two techniques: SMOTE (Synthetic Minority Oversampling Technique) for binary experiments and random resampling for multiclass experiments.

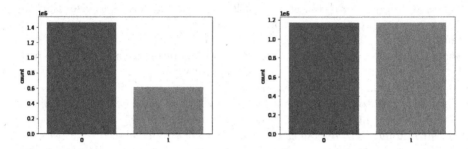

Fig. 5. Class distribution between the DGA & Non-DGA before resampling (a), and after resampling (b).

The big challenge with the dataset was that there was a high-class imbalance and unfair distribution between malware families (see Figs. 5, 6, and 7). In the binary problem, the SMOTE oversampling technique was used to balance the sample between DGA and non-DGA data.

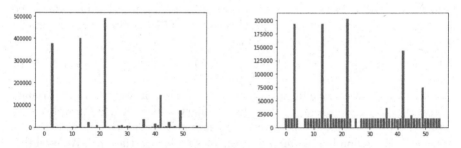

Fig. 6. Class distribution between the malware family before resampling (a), and after resampling (b).

In a multi-class problem, we noticed that the malware families were not distributed properly. Most of these families had a minimal representation in the data, accounting for less than 0.1% of the total data. This led to some DGA families being unavailable in the training set, resulting in poor classification during testing. To address this issue, we removed the domain families whose total instances were less than 5 and applied an appropriate random resampling method to increase the number of minority classes and decrease the number of majority classes to some extent while keeping the total size of the data. To balance the samples, we implemented oversampling techniques on the malware families of training data that accounted for less than 0.5% of the dataset. We also implemented the under-sampling method for domains such as 'legit', 'emotet', and 'rovnix', which had a larger representation in the data.

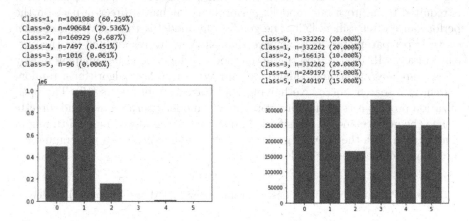

Fig. 7. Class distribution between super-grouped family before resampling (a), and after resampling (b).

3.3.5 Embedding: Embedding is a popular deep-learning technique used to handle categorical data [13]. Word embedding assigns low-dimensional vectors to words [19]. The domain is a sequence of string characters. So, text embedding is used to convert a domain's string of characters to a vector of real values. By performing sentiment analysis, they are integrated into the same procedure. Text embedding can be used only for a model of an input that requires the selection of a text-encoded feature. To represent input tokens, character-level embedding is used. This is a vector representation of characters that is used to handle out-of-vocabulary (OOV) words that do not exist in the trained word vectors. Character-level embedding is particularly useful for extracting morphological information from each word token [20]. It uses character-level features as input to neural networks for natural language processing. To split domain names at a character level, character-level tokenization is used on the labeled benign and malicious domain datasets. Each character in the input domain is mapped to a character vector, which is referred to as a character-level embedding. Then, this is converted into sequences.

3.3.6 Padding: Each sequence has a different length of characters due to varying character lengths among domain families. To address this issue, the sequence of characters is padded with zeros. A padding token is used to obtain a fixed-length input. Padding helps to prevent the feature maps from shrinking and ensures that useful information is not lost.

3.3.7 Hyperparameter Tuning: In machine learning, hyper-parameter tuning is the problem of choosing a set of optimal hyper-parameters for a learning algorithm that is meant to boost a machine learning model [21]. Optimization is required to improve our model's performance. It has a direct impact on the performance of classifiers [22]. The goal of any model is to achieve a minimum error; hyper-parameters help achieve that as they are responsible for the outcome of any ML model. The tuning method can be done either manually or by using searching algorithms. There are four types of search algorithms: random search, grid search, successive halving, and Bayesian optimizers [19]. The optimization process is essential for improving the model's performance and directly affects the performance of classifiers. In our case, we employed manual tuning in combination with the Halving grid search technique to achieve a minimum error for the model's outcome (Table 2).

Table 2. Parameter specifications for training the LSTM & CNN hybrid models.

Parameters	Multiclass		Binary	
	LSTM	CNN	CNN	LSTM
epoches	100	100	30	30
Bach size	1000	1000	1000	1000
Kernel size	–	5	3	–
Filter size	64	64	32	32
activation	'relu'		'relu'	

- Epoch: the number of iterations if the batch size is the entire dataset. It refers to a single, complete pass of the training dataset through the algorithm.
- Bach size: the number of samples that pass through the neural network at a time. As the batch size increases, the training speed increases.
- Kernel: the matrix swiped or convolved across a single channel of an input tensor. The size of the kernel indicates the width and height of the filter.
- Filter size: is the collection of all kernels used in the convolution of channels of the input tensor.
- The activation function defines how the weighted sum of inputs is transformed into an output from neurons. The ReLU activation function is a non-linear function used in many neural networks, especially convolutional neural networks (CNNs) and multilayer perceptions. It solves the vanishing gradients issue by introducing non-linearity to a deep learning model.

3.4   Evaluation Metrics

Once the model has been trained, it's important to evaluate its performance using various metrics to determine how well it performs on new data. Different evaluation metrics, such as accuracy, precision, recall, F1 score, true positive rate, and false positive rate, were utilized to measure the effectiveness of the proposed model.

– Accuracy measures the number of correctly predicted data points out of the total number of data points.

$$Acc = (TP + TN)/(TP + TN + FP + FN) \tag{1}$$

– The F1 score integrates the precision and recall measures, which are regarded as good indicators of the relationship between them.

$$F1Score = 2 * TP/(2 * TP + FP + FN) \tag{2}$$

– True Positive Rate (TPR): proportion of correctly identified positives out of the total actual positives

$$TPR = TP/(TP + FN) \tag{3}$$

– The false positive rate (FPR) is the ratio of false positives to the total number of actual negative events.

$$FPR = FP/(TN + FP) \tag{4}$$

– The Area Under the Curve (AUC) score is the area under the curve, which measures a model's ability to distinguish between classes. It summarizes the overall performance of the model.

4   Results and Discussion

4.1   Binary Experiment

4.1.1   Evaluation of Deep Hybrid Learning Model in Binary Classification.

According to the classification report of the hybrid model, the combination of LSTM with LR, RF, ET, and XGB machine learning models achieved an accuracy rate of 99%. In contrast, the hybrid CNN model resulted in a 98% accuracy score with RF and ET machine learning classifications. The hybrid of LSTM with the XGB model scored the highest AUC at 99.85%, compared to the other models. The hybrid of LSTM with AB obtained the lowest FPR of 0.6%. Additionally, the hybrid model of RF with LSTM achieved the highest TPR of 98.36% along with a 97.7% f1 score. This suggests a higher recall, indicating low FP and FN values (Table 3).

Table 3. Performance of Deep Hybrid Models in binary experiment.

	LSTM					CNN				
	Fl score	TPR (%)	FPR (%)	AUC	Accuracy	Fl score	TPR (%)	FPR (%)	AUC	Accuracy
LR	97.53	98. 19	0.69	99.85	99	94.58	96.23	2.1	99.32	97
NB	96.67	98.04	2.00	99.31	98	90.28	95.99	2	98.17	94
DT	96.65	98.03	2.05	97.99	98	94.24	96.59	3.5	96.52	97
AB	97.33	98.01	0.66	99.83	98	94.78	96.69	2.7	99.39	97
XGB	97.53	98. 19	0.69	99.85	99	95.45	96.74	1.5	99.55	97
RF	97.70	98.36	0.70	99.80	99	96.53	97.43	0.9	99.74	98
ET	97.68	98.35	0.70	99.79	99	96.37	97.73	0.9	99.75	98

While both LSTM and CNN models are capable of learning and representing character features from textual domain names, the hybrid of LSTM with a machine learning algorithm produced better results than the hybrid of CNN in terms of lower FPR (see Fig. 8). This is due to the LSTM model's superior domain name feature representation, as evidenced by its better AUC score, TPR, and lowest FNR. Nonetheless, the performance of CNN is also noteworthy. It is capable of learning and representing character features from textual domain names. In general, we can observe that both CNN and LSTM feature extraction lead to successful results in machine learning algorithm classification. These results are similar and applicable to character representation and feature extraction.

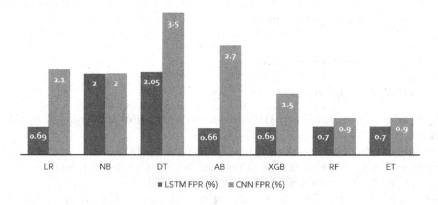

Fig. 8. FPR of DHL model in binary experiment.

4.1.2 Evaluation of Binary Models on Previously Untrained DGA Family.

For the experiment, we tested the deep hybrid learning models on recently collected data from the DomCop and DGArchive datasets. To experiment, we extracted a domain family that is not currently available in NetLab360 or Alexa. The dataset used in the experiment consisted of 18 families, including the legitimate family, and had a total of 1,339,087 data points.

From the test results of the previously unseen malware family, the hybrid model of LSTM with NB achieved the highest AUC score of 80.3% with an f1 score of 86.14%. Additionally, the lowest FPR of 5.4% is observed from the hybrid model of LSTM with the RF in a zero-day malware detection experiment (see Table 4). The result revealed that the LSTM algorithm outperformed the CNN in zero-day malware detection as well.

Table 4. Performance of Hybrid Models on previously unseen domain family.

	LSTM				CNN			
	Fl score	accuracy	AUC	FPR	Fl score	accuracy	AUC	FPR
LR	85.66	77.48	60.3	5.9	84.58	77.32	78.5	13
NB	86.14	79.43	80.3	10.4	73.77	65.14	64.9	31
DT	84.59	76.02	63.9	7.9	81.92	72.84	62.9	14
AB	85.82	77.69	61.2	5.6	85.06	77.97	78.2	12.3
XGB	85.76	77.60	64.2	5.7	85.19	77.43	76.4	9.2
RF	85.83	77.67	62.5	5.4	84.72	76.28	69.8	8
ET	85.81	77.65	61.9	5.5	84.73	76.33	72.9	8.1

4.2 Multiclass Experiment

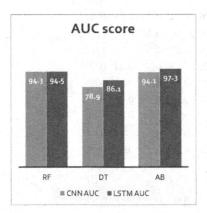
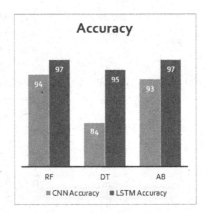

Fig. 9. AUC and Accuracy score of DHL model in multiclass experiment..

Table 5. Performance of Deep Hybrid Models in the multiclass experiment.

DGA Families	CNN			LSTM		
	RF	DT	AB	RF	DT	AB
abcbot	1.00	0.74	0.95	1.00	0.95	1.00
antavmu	1.00	0.71	1.00	1.00	1.00	0.77
bamital	1.00	0.89	0.93	0.93	0.97	0.90
banjori	1.00	1.00	1.00	1.00	1.00	1.00
chinad	0.10	0.16	0.17	0.93	0.89	0.91
cryptolocker	0.00	0.01	0.00	0.05	0.05	0.04
dyre	0.91	0.72	0.84	1.00	0.99	0.99
emotet	0.93	0.79	0.92	0.99	0.98	0.99
enviserv	0.91	0.50	0.77	0.94	0.90	0.93
feodo	0.78	0.37	0.67	0.42	0.37	0.42
flubot	0.00	0.07	0.01	0.93	0.86	0.93
gameover	0.81	0.64	0.79	0.99	0.99	0.99
cre gspy	0.17	0.19	0.27	0.93	0.86	0.90
kfos	1.00	0.95	1.00	0.98	0.94	0.98
legit	0.97	0.92	0.96	0.97	0.95	0.97
locky	0.00	0.01	0.00	0.00	0.01	0.01
matsnu	0.07	0.12	0.10	0.21	0.16	0.20
murofet	0.86	0.60	0.84	0.88	0.73	0.88
mydoom	0.92	0.66	0.89	0.99	0.98	0.99
necro	0.26	0.09	0.22	0.48	0.25	0.47
necurs	0.01	0.03	0.02	0.24	0.19	0.24
ngioweb	0.87	0.70	0.86	0.90	0.82	0.90
nymaim	0.11	0.06	0.10	0.08	0.06	0.08
padcrypt	0.32	0.24	0.36	0.81	0.66	0.71
Pykspa_v1	0.94	0.93	0.92	0.95	0.92	0.95
qadars	0.68	0.45	0.64	0.83	0.78	0.83
ramnit	0.10	0.07	0.13	0.50	0.28	0.48
ranbyus	0.00	0.02	0.00	0.79	0.67	0.76
rovnix	0.98	0.95	0.97	1.00	1.00	1.00
shifu	0.57	0.26	0.49	0.55	0.42	0.55
shiotob	0.72	0.46	0.64	0.87	0.81	0.86
simda	0.88	0.77	0.88	0.94	0.91	0.94
suppobox	0.73	0.49	0.72	0.77	0.69	0.77
symmi	0.40	0.20	0.34	0.63	0.47	0.64
tinba	0.92	0.74	0.87	0.97	0.95	0.97
tofsee	1.00	0.86	1.00	0.92	0.57	0.75
tordwm	0.80	0.56	0.72	0.90	0.81	0.82
virut	0.65	0.44	0.60	0.57	0.44	0.55
Accuracy	94	84	93	97	95	97
weighted average	92	85	91	97	95	97
AUC	94.3	78.9	94.1	94.5	86.1	97.3

4.2.1   Evaluation of Deep Hybrid Learning on Multiclass Classification.

To balance the dataset, the oversampling method was applied to the malware families that are underrepresented (less than 0.5% of the dataset) and the undersampling method to domains that are overrepresented, such as 'legit', 'emotet', and 'rovnix'. This helped to ensure that all classes in the dataset were represented more equally (see Fig. 6).

In our multiclass experiment, we utilized a hybrid of LSTM and CNN models with a machine-learning algorithm. The domain name features were extracted from an intermediate layer of trained LSTM and CNN models. Evaluating multiclass classification using only accuracy is not appropriate since the models may result in high accuracy even if the minority class has bad classification results. This is called the accuracy paradox. The values in the Table 5 represent the f1 score results of the models in detecting the DGA family. The hybrid LSTM with AB model scored 97.33% AUC with 97% f1 score performance, and the hybrid with RF scored 97% accuracy with 94.5% AUC. Various DGA families, including 'bamital', 'banjori', 'dyer', 'emotet', 'flubot','mydoom','symmi', 'tinba', 'tofsee', and 'tordwm', are detected successfully. Out of these, 'chinad', 'gameover', and 'rovenix' are domains that are generated from the combination of numerals and characters, whereas the others are character-based DGA domains. However, some of the family remained undetected, such as dictionary-based DGA (matsnu) and character-based (cryptolocker, nyamin, and locky) domain names.

4.2.2   Evaluation of Hybrid Model on Super-Grouped Multiclass Data.

In the previous experiment, we observed a high misclassification rate of the malware family. This issue can be attributed not only to the algorithm's inability to learn during training but also to the highly imbalanced data and an identical distribution between each family. Moreover, the resampling method employed did not help the situation. Additionally, some DGA-based domains are similar in terms of their character distribution, making classification difficult. To address this challenge, we grouped the data into six supergroups based on the description of the malware family on the Netlab website (see Table 6).

- Group one: legit domain of Alexa.
- Group two: a DGA family generated using random English characters.
- Group three: DGA family generated from the combination of number and character.
- Group four: DGA family generated from a hexadecimal notation.
- Group five: DGA family generated from combined dictionary words.
- Group six: DGA formed from word and hash characters.

Table 6. Super-group of the malware family.

Group	Family	Amount
Group one	Legit	613,032
Group two	'abcbot','banjori','blackhole','xshellghost','virut','vidro','vawtrak','tofsee','tinba'	1251686
	'tempedreve','symmi','simda','shifu','ranbyus','ramnit','Pykspa_v2_real'	
	pykspa_v2_fake','Pykspa_v1','proslikefan','padcrypt','nymaim','necurs','necro',	
	'mydoom','murofet','mirai','locky','fobber_v2','fobber_v1','flubot', 'feodo', 'emotet'	
	'dyre', 'dircrypt', 'cryptolocker', 'conficker', 'blackhole', 'banjori', 'madmax'	
Group three	'shiotob', 'rovnix','qadars','gameover','chinad','ccleaner','tinynuke'	201043
Group four	'omexo', 'gspy','enviserv','copperstealer','bamital','antavmu','tordwm'	1273
Group five	'ngioweb', 'matsnu','bigviktor','suppobox'	9480
Group six	'kfos'	

In the experiment on super-grouped multiclass classification, we used a hybrid model consisting of LSTM and a machine learning algorithm. LSTM achieved the best performance combined with RF and LSTM with AB models for detecting the dictionary-based super-group family, with an F1 score of 84%. In addition, all LSTM hybrid models achieved an F1 score of 98% in detecting legitimate domains, while other groups were also successfully detected (Table 7).

Table 7. Performance of Deep Hybrid Models on the grouped superfamily.

	LSTM					CNN				
	DT	RF	LR	NB	AB	DT	RF	LR	NB	AB
Group1	98	98	98	97	98	95	97	97	86	98
Group2	99	99	99	98	99	98	99	99	96	99
Group3	99	99	99	99	99	97	99	98	98	99
Group4	95	95	95	80	95	82	90	80	25	91
Group5	83	84	82	75	84	57	71	74	15	69
Group6	100	100	98	100	100	100	100	95	93	100
weighted avg	98	99	99	98	99	96	98	98	93	98
Accuracy	98	99	99	98	99	96	98	98	91	98
AUC	97.6	98.9	99.66	99.23	98.92	94.33	99.62	99.65	98.82	99.58

In the experiment for detecting the dictionary-based DGA family, grouping the same type of malware based on their character distribution helped the models understand and generalize their features easily. All supergroups were detected and classified effectively by the hybrid models in this experiment.

4.3 Comparison of Proposed Deep Hybrid Models with Existing Work

Table 8. Binary model comparison with existing work.

Approach	Dataset	Fl score	Accuracy	AUC	TPR (%)	FPR(%)	Reference
Deep learning	Public source	95.56	94.9%	–	–	–	[17]
BILBO	Alexa & DGArchive	96.60	96.56	99.44	95.57	4.54	[14]
LSTM +RF	Netlab360 $ Alexa	97.70	99%	99.80	98.36	0.7	Proposed method

Table 9. Multiclass model comparison with existing work.

DGA Families	Proposed LSTM+RF	LSTM with attention mechanism [15]	LSTM.MI [10]
bamital	0.93	–	0.71
banjori	1.00	0.99	0.99
cryptolocker	0.05	0.35	0.01
dircrypt	0.17	–	0.00
dyre	1.00	1.00	0.98
legit	0.97	0.99	0.98
locky	0.00	0.00	0.00
matsnu	0.21	–	0.00
murofet	0.88	0.74	0.67
necurs	0.24	0.27	0.16
nymaim	0.08	0.17	0.13
padcrypt	0.81	–	0.94
Pykspa_v1	0.95	0.79	0.77
qadars	0.83	–	0.57
ramnit	0.50	0.57	0.69
ranbyus	0.79	0.60	0.52
shifu	0.55	–	0.43
shiotob	0.87	0.94	0.90
simda	0.94	0.94	0.89
suppobox	0.77	– ·	0.14
symmi	0.63	–	0.13
tinba	0.97	0.95	0.93
virut	0.57		0.00

In the binary experiment, the proposed hybrid model of LSTM with an RF algorithm performed better than two other works in terms of five evaluation metrics. The hybrid architecture of LSTM, a CNN, and an ANN named Bilbo in [14] had a higher FPR in the binary experiment; see Table 8. In [17], the deep-learning approach (CNN and RNN) with additional knowledge-based feature extraction achieved low performance for DGA classification due to the smaller amount of training data used (Table 9).

In the multiclass experiment, [15] used an LSTM with an attention mechanism method for DGA domain name classification. On the other hand, [10] presented the LSTM.MI algorithm that used RUSBoost, oversampling, and threshold-moving methods to handle the multiclass imbalance problem. However, their model failed to recognize some of the malware, including the dictionary-based family.

Our proposed LSTM+RF model outperformed the work of [15] and [10] in detecting malware families with better F1 scores. Specifically, in ungrouped multiclass models, we detected thirteen malware families with better F1 scores than [15] and five out of fifteen malware families with better F1 scores than [10]. We used the common families mentioned in these papers for the comparison.

5   Conclusion and Future Work

Manual feature engineering on a dataset has resulted in lower performance and accuracy than classical machine learning methods, according to research [23]. Alternatively, the final classification or clustering layer of a deep learning model driven by fully connected neural network layers may result in overfitting when less data is fed, or it may require unwanted usage of computational resources and power, which does not occur in classical machine learning algorithms. The Deep Hybrid approach (ML and DL) will leverage the benefits of both approaches while alleviating their drawbacks to increase prediction accuracy and decrease computational complexity [23,24]. In this approach, deep learning models were used for feature extraction purposes, while machine learning models were used for the final prediction purpose. This hybrid model is therefore capable of automatically learning features from a given domain. Additionally, we have utilized various techniques to optimize our model, starting from preprocessing techniques to enhance data quality to class balancing techniques such as SMOTE and resampling to prevent model biasing towards the majority class, which can have a significant impact on the final performance of the models. In our binary experiment, utilizing a large amount of training data with a variety of DGA families has allowed our models to achieve improved detection accuracy and even promising results in a novel DGA experiment. In our multiclass experiment, the implemented resampling technique, feature representation using LSTM, and the final classification made by the RF algorithm have contributed to an improved classification result.

We evaluated deep hybrid models for binary and multiclass classification experiments. We applied a combination of LSTM and CNN models with machine learning to the prepared dataset and found that the LSTM model outperformed the CNN model. This suggests that LSTM has a better ability to represent textual features than CNN. In the binary experiment, the hybrid model of LSTM with RF achieved a 99% detection accuracy and a 99.8% AUC score, outperforming other hybrid models. We also tested the models on previously untrained malware families. The hybrid model of LSTM with NB achieved the highest AUC score of 80.3% with an f1 score of 86.14%. Additionally, the lowest FPR of 5.4% is observed from the hybrid model of LSTM with the RF in a zero-day malware detection experiment. In the multiclass experiment, we found that the hybrid models performed poorly in dictionary-based DGA (matsnu) and character-based ('cryptolocker', 'nyamin', and 'locky') domain names. However, when we grouped similar domains based on their character distribution, we achieved better results. The hybrid LSTM model performed better than other models for grouped multiclass classification experiments, achieving a detection accuracy of 99% and an AUC score of 99.65. In this experiment, we detected dictionary-based, character-based, and hexadecimal-based families with f1 scores of 84%, 99%, and 95%, respectively. Grouping the same type of malware helped the models understand and generalize their features easily. Generally, LSTM was found to be an effective model for feature representation and performed well in both binary and multiclass classification tasks. The random forest (RF), which is a forest of decision trees bagged together to form a final prediction, was successful in performing the final classification. The domain name feature representation using the DL algorithm was successful and showed improved performance by combining DL and ML, resulting in a promising result in the Novel DGA family experiment.

Netlab360 contains a highly imbalanced DGA family, and balancing techniques did not significantly improve the detection of some domain families. This resulted in misclassification, especially in an ungrouped multiclass experiment. To improve future results, we recommend training hybrid models on a balanced dataset, equally selecting DGA domain families from DGArchive. Additionally, we recommend extensive training and testing of DHL models for adversarial attacks and working with real-time traffic.

References

1. Robberechts, P., Bosteels, M., Davis, J., Meert, W.: Query log analysis: detecting anomalies in DNS traffic at a TLD resolver. In: Monreale, A., et al. (eds.) ECML PKDD 2018 Workshops. Communications in Computer and Information Science, vol. 967, pp. 55–67. Springer, Cham (2019). https://doi.org/10.1007/978-3-030-14880-5_5
2. Sivaguru, R., Peck, J., Olumofin, F.G., Nascimento, A.C.A., Cock, M.D.: Inline detection of DGA domains using side information. IEEE Access **8**, 141910–141922 (2020)
3. Zago, M., Perez, M.G., Perez, G.M.: A review of scalable detection of botnets based on DGA. In: Proceedings of the Conference Name (2019)
4. Ren, F., Jiang, Z., Wang, X., Liu, J.: A DGA domain names detection modeling method based on integrating an attention mechanism and deep neural network. Cybersecurity **3**, 4 (2020)
5. Ariyapperuma, S., Mitchell, C.: Security vulnerabilities in DNS and DNSSEC. In: Proceedings of the Conference Name, pp. 335–342 (2007)
6. Palau, F., Catania, C.A., Guerra, J., García, S., Rigaki, M.: DNS tunneling: a deep learning based lexicographical detection approach (2020). arXiv:abs/2006.06122

7. Greig, J.: Number of command-and-control servers spiked in 2022: report (2022). www.therecord.media/number-of-command-and-control-servers-spiked-in-2022-report

8. Kumar, S., Bhatia, A.: Detecting domain generation algorithms to prevent DDOS attacks using deep learning. In: Proceedings of the Conference Name (2020)

9. Ravi, V., Alazab, M., Srinivasan, S., Arunachalam, A., Soman, K.P.: Adversarial defense: DGA-based botnets and DNS homographs detection through integrated deep learning. IEEE Trans. Eng. Manage. **70**, 249–266 (2023)

10. Tran, D., Mac, H., Tong, V., Tran, H.A., Nguyen, L.G.: A LSTM based framework for handling multiclass imbalance in DGA botnet detection. Neurocomputing **275**, 2401–2413 (2018)

11. Chowdhury, S.A.: Domain generation algorithm - DGA in malware (2019). www.hackersterminal.com/domain-generation-algorithm-dga-in-malware//

12. Berman, D.S.: DGA CapsNet: 1D application of capsule networks to DGA detection. Information **10**, 157 (2019)

13. Peck, J., et al.: CharBot: a simple and effective method for evading DGA classifiers. IEEE Access **7**, 91759–91771 (2019)

14. Highnam, K., Puzio, D., Luo, S., Jennings, N.R.: Real-time detection of dictionary DGA network traffic using deep learning. SN Comput. Sci. **2** (2020)

15. Qiao, Y., Zhang, B., Zhang, W., Sangaiah, A.K., Wu, H.: DGA domain name classification method based on long short-term memory with attention mechanism. Appl. Sci. **9**, 4205 (2019)

16. Ghosh, I., Kumar, S., Bhatia, A., Vishwakarma, D.K.: Using auxiliary inputs in deep learning models for detecting DGA-based domain names. In: 2021 International Conference on Information Networking (ICOIN), pp. 391–396 (2021)

17. Karunakaran, P.: Deep learning approach to DGA classification for effective cyber security. In: Proceedings of the Conference Name (2021)

18. Qaid, T.S., Mazaar, H., Al-Shamri, M.Y.H., Alqahtani, M.S., Raweh, A.A., Alakwaa, W.: Hybrid deep-learning and machine-learning models for predicting COVID-19. Comput. Intell. Neurosci. **2021** (2021)

19. Cho, M., Ha, J., Park, C., Park, S.: Combinatorial feature embedding based on CNN and LSTM for biomedical named entity recognition. J. Biomed. Inf. **103** (2020)

20. Arora, M., Kansal, V.: Character level embedding with deep convolutional neural network for text normalization of unstructured data for twitter sentiment analysis. Soc. Netw. Anal. Min. **9**, 03 (2019)

21. Ghawi, R., Pfeffer, J.: Efficient hyperparameter tuning with grid search for text categorization using KNN approach with BM25 similarity. Open Comput. Sci. **9**, 160–180 (2019)

22. Yang, L., Shami, A.: On hyperparameter optimization of machine learning algorithms: theory and practice. Neurocomputing **415**, 295–316 (2020)

23. Stampar, M., Fertalj, K.: Applied machine learning in recognition of DGA domain names. Comput. Sci. Inf. Syst. **19**, 205–227 (2022)

24. Jena, B., Saxena, S., Nayak, G.K., Saba, L., Sharma, N., Suri, J.S.: Artificial intelligence-based hybrid deep learning models for image classification: the first narrative review. Comput. Biol. Med. **137**, 104803 (2021)

A Review and Analysis of Cybersecurity Threats and Vulnerabilities, by Development of a Fuzzy Rule-Based Expert System

Matida Churu[1,2]($\boxtimes$), Dewald Blaauw[1], and Bruce Watson[1,2]

[1] Center for AI Research (CAIR), School of Data Science and Computational
Thinking, and Information Science, Stellenbosch University,
Stellenbosch, South Africa
`mvchuru@gmail.com, dnblaauw@sun.ac.za`
[2] Department of Information Science, Stellenbosch University,
Stellenbosch, South Africa

Abstract. Over the past decade, cybersecurity threats and vulnerabilities have significantly increased, primarily due to the widespread adoption of IoT and the expanding use of systems and networks. As technology advances, cyber attackers continually improve their attack methods. Cybersecurity professionals employ the same technologies as cyber attackers for defense purposes. Effectively addressing this challenge requires the development of reliable and comprehensive cybersecurity systems for detection and mitigation. To tackle this issue, a GNS3-Fuzzy Rule-Based Expert System was created, focusing on assessing the risk of each threat over time. The system involved simulating a Local Area Network in GNS3, where attacks were executed using Kali Linux. Throughout the attacks, key metrics such as PC to Server ping time, PC-to-PC ping time, and Download time were recorded and averaged. These metrics were then utilized as inputs and ranges in the fuzzy rule-based expert system. The fuzzy rule-based expert system was developed using the MATLAB software, the fuzzy logic toolbox, and the Simulink tool. The system's output was the risk level associated with different threats. Based on the collected data and the developed system, it was observed that as the PC-to-server time, PC-to-PC time, and download time increase, there is a corresponding elevation in the risk level of the system. Implementing this proposed system provides a dependable and precise solution for detecting the risk level of threats posed to systems.

Keywords: cybersecurity · ping time · download time · fuzzy expert systems

1 Introduction

While the rise of internet-connected gadgets has its benefits, cybercrime's darker side has cast a shadow over our lives [10]. In our open and connected society,

T. G. Debelee et al. (Eds.): PanAfriConAI 2023, CCIS 2069, pp. 151–168, 2024.
https://doi.org/10.1007/978-3-031-57639-3_7

the looming fear of losing our privacy has sparked a lot of discussion. Organizations' abrupt surge in interconnection has both improved our efficiency and exposed us to the risks of cybercrime. We need to be aware of how we interact with the internet. Organizations are simultaneously conquerors and prisoners thanks to the internet [20]. The internet has made it possible for individuals to connect in previously unheard-of ways. Cybercriminals can now take advantage of vulnerabilities that have been developed as a result [2].

Cybersecurity technology is undergoing undisputed evolution, leading to more challenging cybersecurity vulnerabilities [3] and threats. Nearly 40% of states worldwide anticipate cyberattacks, making cybersecurity a global concern that requires integrated efforts at all levels [4]. Data and organizational infrastructure can be protected through various methods, such as intrusion detection, virus prevention, rigorous adherence to good security policies, and many more [1]. Cyber dangers come in various shapes and sizes, including those from terrorists, hackers, and corporate espionage [16]. Even if each assault has a distinct motivation, it should be handled with great care since it puts an organization and its data at risk.

Fuzzy rule-based expert systems have become popular in the detection and management of cybersecurity attacks over the past decade. The authors in [12, 15] and many others have adopted and used fuzzy rule-based expert systems in the realm of cybersecurity. The expert systems previously developed have shown success in the detection and management of cybersecurity, specifically for system administrators. According to [3], the fuzzy set theory's ability to deal with uncertainty and instability connected with information has garnered its popularity in cybersecurity.

Amidst all these, this research aims to develop a cybersecurity threat detection model that enhances the management and risk detection of cyber threats in organizations, governments, and institutions that use systems and or networks. To address these issues the paper proposes a fuzzy rule-based expert system threat model to address some of the threats and vulnerabilities that organizations face.

Due to technological advancements, many daily tasks have been automated since the beginning of the digital era. The recent developments in artificial intelligence have heightened the threat of cyber-attacks, affecting every organization. When these attacks occur, they can result in catastrophic consequences [10]. Fuzzy rule-based expert systems have proven to be effective in identifying online dangers. However, there are still gaps in the detection and management of cyber threats. Therefore, this study proposes a fuzzy rule-based expert system to enhance the management and detection of cyber threats.

This paper seeks to answer the following question: 1. How can fuzzy rule-based expert systems be best applied in the management and detection of cyber threats?

2   Literature Review

2.1   Cybersecurity

A new age of cyber security problems has emerged with the growth of the Internet. New difficulties are being connected with safeguarding information from internal risks, including data breaches and insider theft, in addition to the threat of criminal hackers and foreign governments. For sensitive infrastructures, important assets, and sensitive data, cyber security is a crucial cross-cutting problem [10]. Cybersecurity is a wide phrase that includes any actions done to protect an entity against cyber threats, such as protecting data and minimizing the effects of an event [16]. Understanding the various cyber threats and how to defend against them is crucial. Today's society is seeing an increase in cyberattacks. With the right security precautions, these assaults may be avoided. Since computer networks were widely used, the security of computer and networking systems has become a problem.

2.2   Fuzzy Rule-Based Expert System

In 1965, Zadeh developed the fuzzy set theory [23]. The use of fuzzy set theory has been suggested for three primary study areas, according to [5]: ambiguous phenomena (vague relations in the modeling of issues), information vagueness, and heuristic algorithms. As a result, while analyzing data quantitatively or qualitatively, it is possible to employ linguistic variables instead of precise, crisp values or to provide intervals. As a result, the fuzzy set theory offers more descriptive research [22]. The tendency of decision-makers to see interval judgments as more assured than fixed-value judgments is also crucial [21]. To solve challenging real-world issues, fuzzy rule-based systems (FRBSs) are models built on fuzzy sets first described by [23]. FRBSs convey knowledge in a set of fuzzy rules. Because FRBSs make it possible to deal with uncertainty, imprecision, and non-linearity, the ideas are widely used [19]. Another factor is their interpretability, which enables human specialists to interpret, confirm, and modify the FRBS model created from the data.

2.3   Cybersecurity Threats

Cybercrime involves the use of digital means to perpetrate fraud, steal information, or inflict harm. In essence, unlawful operations that are started via computers, such as hacking, phishing, malware distribution, online stalking, and identity theft, among many others, are collectively referred to as cybercrime [6]. Cybercrime has changed significantly over time and will continue to do so, much like any other type of crime. Some of the threats and vulnerabilities that systems and networks continue to encounter are; denial of service (Dos) attacks, man-in-the-middle, malware, logic bomb, trojan, horse, and network attacks. Distributed Denial of Service (DDoS) and Denial of Service (DoS) attacks have garnered significant concern among researchers due to their rapid proliferation driven by a multitude of attacks and strategies [9,13].

2.4   Fuzzy Logic in Cybersecurity

Cybersecurity and intrusion detection have emerged as a significant field of research because it is not theoretically possible to set up a complete system with no fault [15]. Fuzzy rule-based expert systems have been used in cybersecurity in the detection of cybersecurity threats and vulnerabilities. Fuzzy Rule-based expert systems for cybersecurity are systems that consist of pools of rules and a mechanism for accessing and running the principles [7]. In addition to being able to handle uncertainty in intrusion detection, fuzzy rule-based systems, which integrate fuzzy logic with expert system methodology, also enable the most flexible reasoning about the broadest range of available information [11]. Fuzzy rule-based expert systems are very useful in detecting threats and not necessarily eliminating the threats [12,15]. Researchers have employed fuzzy rule-based expert systems to determine their value as cyber indicators that alert system administrators to impending cyber threats [12,14,15,18] The warning signals are generated by the rules embedded in the expert systems [11].

2.5   Time in Cybersecurity

Ensuring cybersecurity is a critical concern for organizations employing numerous IT devices or managing a substantial array of IT assets. Assessing an organization's cyber risk and its ability to withstand cyber-attacks poses significant challenges. Time stands as a pivotal factor in cybersecurity risk assessment due to its correlation with security time the duration a system remains secure before succumbing to successful exploitation [24]. [24] work presents a mathematical framework that employs stochastic modeling techniques in conjunction with Markov chains and attack graphs. This framework aids organizations in analyzing their time resilience against cyber-attacks, enabling the formulation of security strategies geared towards optimal time-based resilience.

In [24] study, resilience is evaluated concerning both the frequency of attacks and the cumulative duration of these attacks. Time also plays a crucial role in botnet detection methodologies. [17] proposed a method centered on observing network traffic behaviors from sampled network packets within specific time intervals to identify potential botnets. Similarly, [8] introduced a two-stage detection approach employing supervised and unsupervised machine learning techniques. Their method aims to differentiate between botnet and non-botnet network traffic by analyzing network flow records collected within defined time intervals.

3   Method

The method proposes a threat model that focuses on detecting the risk imposed on a system with time as the main criterion. Threats were included as input in determining the risk imposed on networks or systems. The threat model was proposed by using GNS3 and fuzzy rule-based expert systems which were implemented and developed in MATLAB. The methodology was done in two steps the first step was collecting data using the GNS3 tools followed by the development of the fuzzy rule-based expert system in MATLAB.

3.1   Data Collection in GNS3

A hypothetical LAN network topology was simulated in GNS3, running on a VMWare virtual machine, which is shown in Fig. 1. A virtual Kali Linux machine was connected to the system to simulate the role of an attacker.

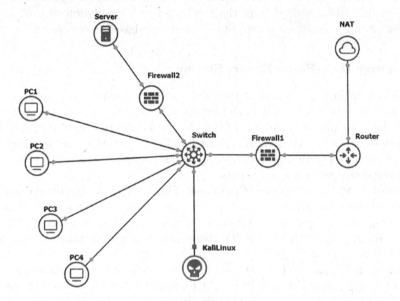

Fig. 1. Hypothetical LAN star network system simulated in GNS3

A LAN star network topology setup was followed, where all devices were connected to a switch that connected the devices to the internet cloud. The network shown in Fig. 1 was the one whose regular traffic and time logs were examined while it operated under normal conditions without attacks, and from which attacks were launched. The assaults targeted the network's important devices, primarily the switch, and throughout the attacks, the system's operation was investigated to determine how the attacks compromised the system or device and whether there were any discrepancies between how the systems performed normally and when under attack. The network was set up with a Dynamic Host Protocol (DHCP), and it was confirmed that all components were assigned unique IP addresses and could communicate with the network and the internet.

Network Analysis in GNS3. Two analyses were done one when the network was operating under normal conditions and one when the network was under attack. It is important to note that the main criterion of the data analysis was based on time. The analysis was based on the following tasks:

i) Pinging between PCs and the server and analyzing the time on each ping.
ii) Pinging between the three PC?s was also done.
iii) Downloading a file from the server was done on each and the min and max time of each ping was recorded, and an average min and max time was recognized.

Based on these tasks, the subsequent data (time) was gathered from the regular network analysis and during the various attacks. Table 1 presents a summary of the data acquired from the GNS3 network simulation utilized in the implementation and development of the fuzzy rule-based expert system.

3.2   Fuzzy Rule-Based Expert System

The first step in the creation of the fuzzy rule-based expert system involves a collection of data. The threat model expert system in this paper was developed using the data and key variables from the data collected from the simulated network from the GNS3 simulation. Based on the information that was captured from the previous section the following variables (inputs and outputs) were found to be respective criteria for this paper.

Input 1: PCtoServerTime (PtoSTime) This criterion evaluates the average round-trip ping time between a PC and the Server. It assesses whether the processing time for pinging between a PC and the server has changed, either slowing down or becoming faster. The three membership functions for this input are low, medium, and high.

Input 2: PC to PC Time (PtoPTime) This criterion measures the average round-trip ping time between PCs. It determines if there have been changes in the processing time for pinging between PCs, such as slowdowns or increases. Similar to PtoSTime, PtoPTime has three membership functions: low, medium, and high.

Input 3: Download Time (DTime) This criterion focuses on the average time required to download files from the server using a PC. Like PtoSTime and PtoPTime, DTime is categorized into three membership functions: low, medium, and high.

Input 4: Threat (T) This criterion identifies various techniques that cyber attackers can employ for cyber crimes. The threats include: Normal, Ping of death, CAM overflow, ICMP, ARP, and DHCP.

Output 1: Risk Level (risk level) The output of the system is the determination of risk levels associated with each threat on the network. Assessing the risk levels of these threats is crucial for systems and networks to detect and mitigate potential risks effectively. The risk levels are classified as minimal, significant, and extreme.

In summary, the threat model architecture analyses inputs related to ping times, download time, and threat types, and produces an output that quantifies the risk levels associated with each threat in the network.

Table 1. Summary of the data acquired from the GNS3 network simulation

Attack	Task	Average time recorded (seconds)		
		MIN	AVG	MAX
Operating under normal conditions	1. PC to Server	0.96067	1.777	2.679
	2. PC to PC	2.181	3.87475	5.59325
	3. Downloading file from server	58,625		
ARP	1. PC to Server	10.0955	16.086	38.89333
	2. PC to PC	10.274	17.2873	41.8643
	3. Downloading file from server	197,325		
Ping of Death	1. PC to Server	1,708	2,6533	4.47433
	2. PC to PC	2.8165	5.0965	9.43125
	3. Downloading file from server	48.25		
CAM Overflow	1. PC to Server	1.61817	3.630167	7.481467
	2. PC to PC	4.119	7.346	14.237
	3. Downloading file from server	34.125		
DHCP attack	1. PC to Server	1.737833	13.42433	61.70883
	2. PC to PC	2.628	7.92575	28.50775
	3. Downloading file from server	File did not download		
ICMP attack	1. PC to Server	1.964167	4.384667	12.87817
	2. PC to PC	3.4195	6.508	15.295
	3. Downloading file from server	23.875		

Architecture of the System. The system architecture comprises four inputs and one output, with three inputs related to time and the remaining input indicating the threat level faced by a system or network. The output of the threat model is dedicated to assessing the risk level of the threat, taking into consideration the processing time for each task. This analysis relies on the previously captured and calculated time data. The systems architecture is shown below in Fig. 2. Figure 2 depicts the system's four inputs and output, which were explained above in the previous section.

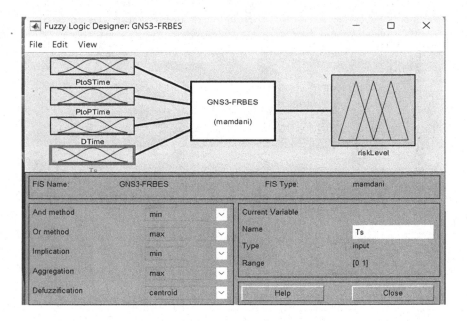

Fig. 2. Architecture of Fuzzy Rule-Based Expert System for cybersecurity

Fuzzy Rule-Based Expert System Rules Creation. Twenty rules were derived based on the four inputs and one output. The system rules main goal is based on time which is based on the data collected and captured in the first section of the method using the GNS3 tool. Table 2 provides a comprehensive overview of the system's architecture. It summarizes essential details such as the model type, the number of inputs and outputs, and the scales associated with different inputs and outputs, offering a concise understanding of the system's structure.

The rules aim to ensure that if the processing time of a certain task in the network or system surpasses a certain time or if the processing time is low or high or abnormal the system alerts the users and system administrators of the risk level associated with that specific threat. The rules were established by analyzing

Table 2. GNS3-FRBES System Architecture

Category	Parameter	Value	Description
Type	Type	Mamdani	Mamdani Fuzzy Logic
Time and Risk Level	Inputs	[4 1]	Input Variables
	NumInputMFs	[3 3 3 6]	Number of Input Membership Functions
	NumOutputMFs	[1]	Number of Output Membership Functions
	AndMethod	min	And Method
	OrMethod	max	Or Method
	DefuzzMethod	centroid	Defuzzification Method
Input Labels	PtoSTime	[0 25]	Input Label and Range
	PtPTime	[0 22]	Input Label and Range
	DTime	[0 1]	Input Label and Range
	Ts	[0 1]	Input Label and Range
Output Labels	RiskLevel	[0 1]	Output Label and Range

the data collected from GNS3, with time as the primary factor in determining these rules. By examining the duration of each threat or attack on the system, the rules were formulated. The findings from GNS3 revealed that distributed denial of service attacks, specifically ICMP and DHCP flood attacks, had the most severe and harmful impact on the system's ping and download time. Similarly, man-in-the-middle attacks, particularly ARP attacks, also significantly affected the ping and download speeds of the system. On the other hand, denial of service attacks such as ping of death and CAM overflow attacks had minimal to no effect on the system's ping and processing times. Therefore, time played a crucial role in determining these rules, as illustrated in Fig. 3.

1. If (PtoSTime is low) and (PtoPTime is low) and (DTime is medium) and (Ts is normal) then (riskLevel is minimal) (1)
2. If (PtoSTime is low) and (PtoPTime is low) and (DTime is low) and (Ts is pingOfDeath) then (riskLevel is minimal) (1)
3. If (PtoSTime is medium) and (PtoPTime is medium) and (DTime is medium) and (Ts is pingOfDeath) then (riskLevel is significant) (1)
4. If (PtoSTime is low) and (PtoPTime is medium) and (DTime is low) and (Ts is camOverFlow) then (riskLevel is minimal) (1)
5. If (PtoSTime is medium) and (PtoPTime is high) and (DTime is medium) and (Ts is camOverFlow) then (riskLevel is significant) (1)
6. If (PtoSTime is low) and (PtoPTime is low) and (DTime is low) and (Ts is ICMP) then (riskLevel is significant) (1)
7. If (PtoSTime is medium) and (PtoPTime is medium) and (DTime is medium) and (Ts is ICMP) then (riskLevel is significant) (1)
8. If (PtoSTime is medium) and (PtoPTime is high) and (DTime is medium) and (Ts is ICMP) then (riskLevel is extreme) (1)
9. If (PtoSTime is medium) and (PtoPTime is medium) and (DTime is high) and (Ts is ARP) then (riskLevel is significant) (1)
10. If (PtoSTime is high) and (PtoPTime is medium) and (DTime is high) and (Ts is ARP) then (riskLevel is extreme) (1)
11. If (PtoSTime is low) and (PtoPTime is low) and (DTime is low) and (Ts is DHCP) then (riskLevel is significant) (1)
12. If (PtoSTime is medium) and (PtoPTime is medium) and (Ts is DHCP) then (riskLevel is significant) (1)
13. If (PtoSTime is high) and (PtoPTime is high) and (Ts is DHCP) then (riskLevel is extreme) (1)
14. If (PtoSTime is high) and (PtoPTime is high) and (DTime is high) then (riskLevel is extreme) (1)
15. If (PtoSTime is medium) and (PtoPTime is medium) and (DTime is medium) then (riskLevel is significant) (1)
16. If (PtoSTime is low) and (PtoPTime is low) and (DTime is low) then (riskLevel is minimal) (1)
17. If (PtoSTime is high) and (PtoPTime is high) then (riskLevel is extreme) (1)
18. If (PtoSTime is medium) and (PtoPTime is high) and (DTime is low) then (riskLevel is significant) (1)
19. If (DTime is high) and (Ts is ARP) then (riskLevel is extreme) (1)
20. If (PtoSTime is high) and (Ts is DHCP) then (riskLevel is extreme) (1)

Fig. 3. List rules of the Fuzzy rule-based expert system

4   Analysis and Results

4.1   Evaluation of Fuzzy Rule-Based Expert Systems

For analysis two sample simulations are used, the first is when the network is under normal conditions and the second is when the network is when the network is under a specific attack. The analysis of these two simulations will help in providing more insight into the systems' capability to detect the risks posed on systems.

Evaluation of System Under Normal Conditions. For analysis when the simulation is under normal conditions the following sample simulation variables are used. The simulation scenarios involve when the attack is under normal conditions. The crisp for the various inputs are as follows; PC-to-Sever time (PtoSTime) is low, PC-to-PC time (PtoSTime) is also low and download time (DTime) is also low. The crisp input values for the sample scenario are:

PtoSTime (high) = 1.77
PtoPTime (high) = 2.28
DTime (low) = 0.457

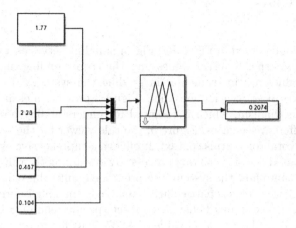

Fig. 4. Simulink of Sample Simulation 1

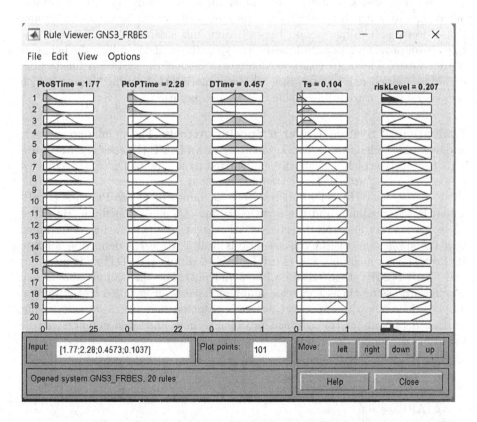

Fig. 5. Sample Solution of Rule Viewer of Simulation 1

Ts (DHCP) = 0.104
riskLevel (extreme) = 0.207

In the system depicted in Fig. 4 and Fig. 5, Simulink serves as a powerful tool for users to input specific crisp values into the corresponding input variables. These input values are instrumental in guiding the system's decision-making process and directly impact its operation. Figure 3 provides a clear visualization of this input mechanism, which is essential for understanding how the system responds to different scenarios. Figure 5, the rule viewer for the sample simulation, plays a pivotal role in this context. It offers a comprehensive overview of the system's rule-based decision-making process. By evaluating the rule viewer, users can gain insight into how the system interprets the input values and assesses the associated risk levels. This is particularly valuable for making informed decisions and fine-tuning the system's behavior to meet specific requirements. The crux of the matter lies in the risk level value of 0.207. This figure represents the system's evaluation of its state when presented with the specified input values. A risk level of 0.207 is indicative of minimal risk, signifying that the system's operation is considered safe and reliable under these conditions. This is a favorable outcome, indicating that the system's responses align with the desired minimal-risk objectives. The interaction between Simulink, crisp input values, and the rule viewer provides a powerful means for evaluating and controlling the behavior of the system. The 0.207 risk level demonstrates that the system is functioning within a minimal-risk framework when encountering the specified inputs, which is essential for ensuring its reliable and secure operation.

Evaluation of System Under a Specific Attack. The simulation scenario involves an attack on the network, identified as an ICMP (Internet Control Message Protocol) attack, specifically categorized as a type of DDoS (Distributed Denial of Service) attack. The system has assessed the parameters as follows: the PC-to-Server time (PtoSTime) is rated as medium, the PC-to-PC time (PtoPTime) is also medium, and the download time (DTime) is medium. Based on these parameters and the detected ICMP threat, the system has determined that the applicable rule for this scenario is Rule 7. Rule 7 is defined as follows: If (PtoSTime is medium) and (PtoPTime is medium) and (DTime is medium) and (Ts is ICMP), then (riskLevel is significant). For this sample simulation, specific processing times and scale numbers have been assigned to the inputs. These values help determine the system's response and risk level assessment. Further details about the assigned processing times and scale numbers can be found in the simulation results:

- PtoSTime (medium) = 12.88
- PtoPTime (medium) = 15.30
- DTime (medium) = 0,24
- Ts (ARP) = 0.6
- riskLevel (significant) = 0.533

The numbers had to be rounded to 2 decimal places due to Simulink and MATLAB not accepting numbers with more than 2 decimal places. After inserting the different times into the Simulink tool in MATLAB the risk level of rule 7 based on data that was proposed above was indicated to be 0.533. This is indicated in Fig. 6 and Fig. 7.

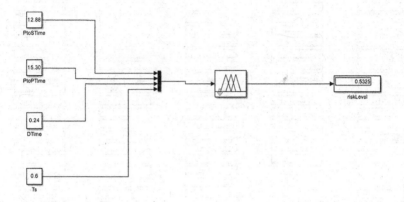

Fig. 6. Simulink for when the system is under a specific attack

By inputting the corresponding values into the Simulink tool, using the provided sample solution, the risk level for the given information was determined to be 0.533, indicating a significant level of risk. This output demonstrates the system's assessment of the potential threat and highlights the need for immediate attention, as there may be disruptions impacting the system's functionality.

To delve deeper into the simulation and gain a more detailed understanding, Fig. 7 presents an expanded view of the sample simulation. The figure allows users to examine the inner workings of the system more closely. In alignment with the previously inserted inputs, such as (PtoSTime = 12.88), (PtoPTime = 15.30), and (Ts = 0.6), the output (risk level) is determined to be 0.6. The rule viewer depicted in Fig. 7 visually depicts how each input affects the corresponding rules within the system. The final rule, represented by the red line, showcases the cumulative impact of all the inputs on the overall risk level. With a risk value of 0.533, it becomes evident that the threat is significant and requires immediate attention to address the disruption to the system's functioning.

A risk level of 0.533 indicates that the threat imposed on the system is at a high level, necessitating a significant and immediate response. The assigned risk level serves as an important metric in determining the severity and potential impact of the threat on the system's security and functionality.

With a risk level of 0.533, it becomes evident that the threat has surpassed a threshold and poses a considerable risk to the system. This signifies that the system's vulnerability has been compromised, and there is a high likelihood of potential damage, disruption, or unauthorized access. In response to such a substantial risk, it is imperative to implement robust security measures, mitigation strategies, and incident response protocols.

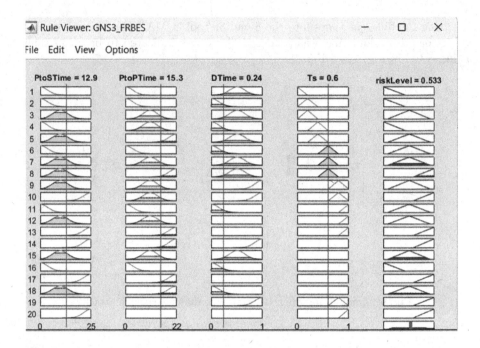

Fig. 7. Sample Solution Rule Viewer of attack

The significance of a risk level of 0.533 cannot be understated. It serves as an alert to the system administrators and stakeholders that immediate action is required to mitigate the potential consequences. The higher the risk level, the more urgent and critical the response becomes. This may involve deploying additional security controls, enhancing network monitoring, isolating affected components, or even initiating incident response procedures.

Expert System Surface Analysis. Utilizing the surface view functionality in MATLAB provides a visual representation of the system's behavior. This visual depiction facilitates the observation of changing dynamics, the identification of patterns and trends, and a deeper understanding of the simulated data. Through this analysis, the system's characteristics can be explored, and assess how the risk level can be assessed as time increases.

PC-to-Server Time and PC-to-PC Time and Risk Level
The analysis of the figure above clearly demonstrates a direct relationship between the increase in PC-to-Server time and PC-to-PC time and the corresponding rise in the risk level. This indicates that, based on the defined rules and membership functions from the previous chapter, an increase in time has a detrimental effect on the system. Notably, Fig. 8 above shows that as time increases in either PC-to-Server time or PC-to-PC Time, the risk level also increases. However, PC-to-PC Time has a more pronounced detrimental effect on the risk level

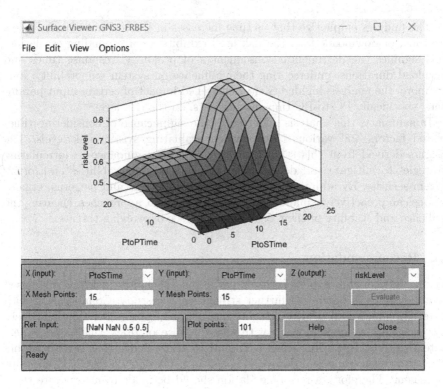

Fig. 8. PC-to-Server Time, PC-to-PC Time and Risk Level Relationship

compared to PC-to-Server time. The surface view of the ICMP attack sample solution depicted in Fig. 5 reveals that when PC-to-Server time and PC-to-PC time are in the medium membership functions, the risk level reaches a significant risk level of around 0.53.

The analysis highlights a clear correlation between time parameters and the associated risk level. The detrimental impact of increasing PC-to-Server time and PC-to-PC time on the risk level is evident, with PC-to-PC time having a more substantial influence. The surface view in Figure provides visual confirmation of how the risk level escalates as both PC-to-Server time and PC-to-PC time increase, particularly when these inputs fall within the medium membership functions.

5 Conclusion

In conclusion, the comprehensive analysis of the sample simulation, rule viewer, and utilization of various surface views in MATLAB have yielded valuable insights into the behavior and risk level of the system when evaluating the impact of different threats. The examination of diverse input parameters, including PC-to-Server time, PC-to-PC time, download Time, and Ts, has uncovered distinct relationships with the risk level.

The findings emphasize that as time increases in PC-to-Server time, PC-to-PC time, or download time, the risk level exhibits a corresponding escalation. This signifies the detrimental consequences of prolonged response times and download durations, underscoring their influence on system vulnerability. Furthermore, the analysis highlights the heightened impact of certain input parameters, specifically PC-to-PC time, and threats, on the risk level.

In summary, this analysis underscores the significance of considering time-related factors and various threats when evaluating system risk levels. The insights derived from this study can provide crucial guidance in formulating strategies to mitigate and detect cyber threats and vulnerabilities and fortify system security. By addressing the identified correlations and patterns, organizations can proactively enhance their risk management approaches, ensuring the resilience and stability of their systems in the face of evolving threats.

5.1 Recommendation

Seeking advice from cybersecurity experts and specialists in fuzzy rule-based expert systems is crucial to further strengthen the study. Collaborating with these individuals will provide insights into backend processing time and facilitate the improvement of rules with the guidance of fuzzy rule-based expert system experts. Regarding time, it is vital to acknowledge that several factors affect processing time and download speed in networks, which GNS3 does not implement. Therefore, a recommendation should be made to incorporate these factors. For instance, network size, including the number of connected devices, plays a significant role. Additionally, internet bandwidth also affects processing time. Considering these factors is crucial.

Recognizing that time can vary among organizations, departments, and networks is crucial for the effective detection and mitigation of cybersecurity threats. Customizing the system according to the rules and regulations of each organization and department in terms of time and speed is necessary. Consequently, the system cannot be universally implemented in all organizations. Instead, it needs to be tailored to align with the specific rules and regulations of each organization and department.

5.2 Future Work

While the hypothetical network simulation provided a solid foundation for assessing and analyzing threats and vulnerabilities to systems, it is important to acknowledge that GNS3, being open-source software, lacks specific real-world hardware, software, and LAN/network structures. To address this limitation, it is imperative to replicate the testing in a simulated emulator capable of mimicking real-world networks and incorporating additional components that organizations commonly employ to protect and mitigate threats, which are not available in the GNS3 simulator.

References

1. Abdymanapov, S., Muratbekov, M., Altynbek, S., Barlybaye, A.: Fuzzy expert system of information security risk assessment on the example of analysis learning management systems (2021)
2. Ahsan, M., et al.: Cybersecurity threats and their mitigation approaches using machine learning- a review. J. Cybersecur. Priv **2**, 527–555 (2022)
3. Alali, M., et al.: Improving risk assessment of cyber security using fuzzy logic inference system. Comput. Secur. **74**, 323–339 (2017)
4. Amna, A., Raul, V.: Cybercrime prevention in the kingdom of Bahrain via IR security audit plans. J. Theor. Appl. Inf. Technol. **65**, 274–292 (2014)
5. Cai, K.: System failure engineering and fuzzy methodology: an introductory overview. Fuzzy Sets Syst. **83**, 113–133 (1996)
6. Chakraborty, A., Biswas, A., Khan, A.K.: Artificial intelligence for cybersecurity: threats, attacks and mitigation. Computer Science>Cryptograph and Security (2022)
7. Chauhan, K.: Fuzzy approach for designing security framework, pp. 173–195 (2021)
8. Ding, S., Bunn, J.: Machine learning for cybersecurity: network-based botnet detection using time-limited flows. California Institute of Technology (2017)
9. Douligeris, C., Mitrokotsa, A.: DDoS attacks and defense mechanisms: classification and state-of-the-art. IEEE Comput. Netw. **44**, 643–666 (2004)
10. Feng, B., et al.: Stopping the cyberattack in the early stage: assessing the security risks of social network users. Security and Communication Networks (2019)
11. Gao, M., Zhou, M.: Fuzzy intrusion detection based on fuzzy reasoning petri nets, pp. 1272–1277 (2003)
12. Goztepe, K.: Designing a fuzzy rule-based expert system for cyber security. Int. J. Inf. Secur. Sci. **1**, 13–19 (2015)
13. Mirkovic, J., Reiher, P.: A taxonomy of DDoS attack and DDoS defense mechanisms. ACM SIGCOMM Comput. Commun. Rev. **34**(2), 39–54 (2004)
14. Mlakic, D., Majdandzic, L.: Fuzzy rule based expert system for SCADA cyber security (2016)
15. Mudassar, M., Kankale, P.A., Gawande, P.: Computing the impact of security attack on network using fuzzy logic. Int. Res. J. Eng. Technol. (2016)
16. Obotivere, B., Nwaezeigwe, A.: Cybersecurity threats on the internet and possible solutions. IJARCEE **9**, 92–97 (2020)
17. Riyaz, B., Ganapathy, S.: An intelligent fuzzy rule-based feature selection for effective intrusion detection. In: International Conference on Recent Trends in Advance Computing (ICRTAC), pp. 206–211 (2018)
18. Riyaz, B., Ganapathy, S.: An intelligent fuzzy rule-based feature selection for effective intrusion detection, pp. 206–211 (2018)
19. Riza, L.S., Bergmeir, C., Herrera, F., Benitez, J.: Fuzzy rule-based systems for classification and regression tasks (2019)
20. Shrestha, J.M., Noll, C., Roverso, J., Davide, A.: A methodology for security classification applied to smart grid infrastructures. Int. J. Crit. Infrastruct. Prot. **28**, 100–342 (2020)
21. Tubis, A., et al.: Cyber-attacks risk analysis method for different levels of automation of mining processes in mines based on fuzzy theory use. Sensors **20**, 7210 (2020)

22. Yasli, F., Bolat, B.: A risk analysis model for mining accidents using a fuzzy approach based on fault tree analysis. J. Enterp. Inf. Manag. **31**, 577–594 (2018)
23. Zadeh, L.: Fuzzy sets. Inf. Control **8**(3), 338–353 (1965)
24. Zhang, Y.: Optimization-time analysis for cybersecurity. IEEE Trans. Dependable Secure Comput. **19**(4), 2365–2383 (2022)

Autonomous Vehicles

Neural Network Based Model Reference Adaptive Control of Quadrotor UAV for Precision Agriculture

Muluken Menebo$^{(\boxtimes)}$ ⓘ, Lebsework Negash ⓘ, and Dereje Shiferaw

Addis Ababa University, Po. Box 385 Addis Ababa, Ethiopia
menebomuluken@gmail.com

Abstract. This paper aims to design a neural network-based model reference intelligent adaptive control for quadrotor UAV and implement a machine learning approach to recognize the severity level of yellow wheat rust for precision agriculture. Yellow wheat rust is a fungal disease that can cause massive destruction in wheat production and quality. Obtaining accurate data from large-scale crops and detecting those diseases based on specific standards via visual inspection become labor-intensive, time-consuming, and sensitive to human error. Addressing these issues involves deploying a quadrotor for data acquisition and training a cutting-edge Convolutional Neural Network (CNN) for image analysis. Since existing control techniques are computationally intensive and show poor performance in tolerating unmatched uncertainty, the proposed controller is designed by using a nested control approach. In this control architecture, feedforward neural networks are trained to estimate position controller parameters online, whereas recurrent neural networks are trained to estimate the model and control the attitude of the quadrotor. Then, Xception CNN is trained by using a transfer learning approach. To verify controller performance, numerical simulations have been conducted in various scenarios. The results show that the designed controller has high tracking precision, robustness, and enhanced antidisturbance ability in nominal scenarios and the presence of matched and unmatched uncertainties, and the retrained model achieves an accuracy of 97.28%. Therefore, the suggested controller is a promising quadrotor control technique, and the retrained Xception model can be used for detecting the severity level of yellow wheat rust.

Keywords: Intelligent Flight Control · Deep Learning · Quadrotor UAV

1 Introduction

The exponential rise in world population calls for a rapid increase in food with the application of available technologies that reduce laborious work, maximize productivity, and reduce side effects on the environment. Precision Agriculture (PA) is a farming management technique that focuses on observing and responding to intrafield variations in the crop. It gained popularity after the realization

T. G. Debelee et al. (Eds.): PanAfriConAI 2023, CCIS 2069, pp. 171–193, 2024.
https://doi.org/10.1007/978-3-031-57639-3_8

of diverse fields of land hold different properties: soil type variability, moisture content, nutrient availability, and so on. By using appropriate data acquisition techniques, gathering such kind of data from the field helps farmers to precisely determine what inputs to put exactly where and in what quantities. This can minimize the cost of inputs while maximizing the productivity, quality, and profit margin [3, 19]. To gather and use information effectively, familiarizing PA with available technological tools is very important since they can automate and simplify the collection and analysis of information. At large scale, precise monitoring of the field is quite a challenging task, so this study mainly focuses on deploying Unmanned Aerial Vehicles (UAV) for data acquisition and implementing a machine-learning approach to detect yellow wheat rust.

Wheat is one of the strategic crops that feed the exponentially growing world population [10]. However, crop diseases caused by viruses, Bacteria, and fungi cause huge destruction not only in the yield but also in quality as well. One of the application areas of PA is detecting those crop diseases at an early stage and taking them under control before they cause huge destruction in the field. Yellow wheat rust is a highly destructive fungal disease of wheat that can damage the entire crop. Research shows that this wheat rust has been seen in more than 60 countries worldwide, and it has the potential to damage 50% to 100% of the crop if it's not controlled and prevented at an early stage [15]. Nowadays, this wheat disease has become a major headache for both farmers and plant pathologists. This is due to difficulties in obtaining accurate data about the whole crop, and sometimes it's challenging to detect such kind plant diseases via visual inspection without using laboratory material. Most of the time wheat disease is inspected manually by humans, so for large-scale crops, it becomes labor-intensive, time-consuming, and sensitive to human error. Therefore, it's critical to find effective and efficient ways that help to recognize yellow wheat rust at an early stage and keep it under control to minimize economic losses. The proposed techniques in this study, allow farmers to monitor their crop condition through the air constantly, and to find problems quickly and accurately by reducing time-consuming activities in ground-level spot check by humans.

UAVs are a kind of recent-generation autopilot aircraft that can be controlled manually by a remote operator or autonomously through pre-programmed flight paths. The quadrotor is a type of UAV that is driven by four rotors with the capability of Vertical Takeoff and Landing [17]. Due to low operation cost and maneuverability at low altitudes, quadrotor application areas have been growing enormously from time to time in various industry, civilian, and military applications. One of the application areas of quadrotors that this study focuses on is in PA. By capturing images from the crop with a built-in camera platform on the body of the quadrotor, providing multiple images for farmers. These help farmers to process images and monitor crops' health while identifying areas of the crops that require specific forms of attention [8, 18]. Since the captured image through the air needs to be accurate, developing an effective controller that has good performance in tolerating parameter variation and external disturbance from the wind is vital for the performance and application of quadrotors in PA.

Adaptive control is a nonlinear conventional control technique that can adjust its parameters online in a dynamic environment by observing the process under control. It is broadly classified into direct adaptive control in which its parameter is adjusted directly, and indirect adaptive control in which parameter estimation and control are performed simultaneously [6]. Currently, adaptive control is mostly used conventional control approach in flight control of UAVs. This is because of its ability to adjust controller parameters in a way that it can be automatically adapted to the changes in the flight dynamics. However, they lose their performance when the system is affected by unmatched uncertainties. In this study, intelligent control is designed to deal with complex systems, which are difficult to be controlled by conventional control techniques. Intelligent control systems combine artificial intelligence (AI) techniques with conventional control approaches in designing autonomous systems [5] which have been showing a huge change in existing control systems. Such systems can achieve sustained behavior under the conditions of matched and unmatched uncertainties due to uncertainty in system dynamics, unpredictable environmental changes, unreliable sensor information, and actuator malfunction [21]. Among different types of AI techniques, the learning capability of Artificial Neural Networks (ANNs) makes them widely used AI techniques in the control engineering arena as black-box estimators and controllers. Furthermore, CNN is a special type of deep ANN that is considered state-of-the-art for real-time image analysis due to its effective results in processing visual data.

To get accurate data from the crop and to make good management decisions, this study aims to develop an intelligent flight control algorithm that combines the learning power of ANN with conventional adaptive control for quadrotor control and train a deep learning model to detect the severity level of yellow wheat rust remotely for PA with the following key contributions:

1. Development of a more precise quaternion-based singularity-free mathematical model of quadrotor flight dynamics,
2. Implementation of ANN-based novel intelligent quadrotor flight control approach,
3. Implementation of online learning algorithm to enhance controller performance, and
4. Proposing and simulating efficient data accusation and analysis mechanism for PA.

The paper is organized as follows. The detailed quaternion-based mathematical model of quadrotor flight dynamics is presented in Sect. 2. ANN-based intelligent flight control design is presented in Sect. 3. In Sect. 4, simulation and training results are presented. Section 5 concludes the manuscript.

2 Modeling Quadrotor Flight Dynamics

2.1 Preliminary Notions of Quadrotor UAV

To model quadrotor flight dynamics, the coordinate frames of the quadrotor must be defined first. In Fig. 1 there are two coordinate frames that the quadrotor will operate in: inertial frame (x_i, y_i, z_i) and body frame (x_b, y_b, z_b) [16]. Inertial Frame(IF) is defined w.r.t the ground with a positive z-axis pointing in the opposite direction of gravity, whereas Body Frame(BF) is defined w.r.t Center of Gravity (CoG) of the quadrotor with its axes fixed to the body.

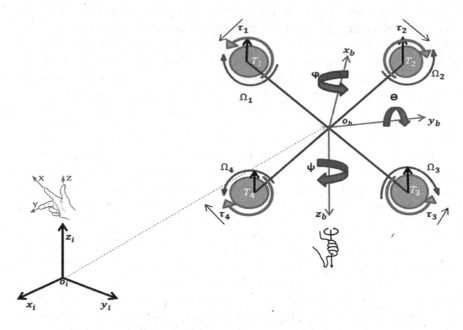

Fig. 1. Quadrotor reference frames and its free body diagram

The necessity of diffing those reference frames is because actuator inputs and thrusts are manipulated w.r.t BF, Inertial Measurement Unit (IMU) measure quantities w.r.t BF, GPS measures position w.r.t IF, and Newton laws are valid in IF only. Hence, coordinate frame transformation is used to transform quantities between the two reference frames. To avoid the nonlinearity, geometric singularity, and computational cost usually connected with the Euler angles [12,13], the quaternion approach is implemented for coordinate frames transformation. Quaternion contains a scalar part q_o, and a three dimensional vector part: $q_v = [qi, qj, qk]$ with orthonormal base $[i, j, k]$ that satisfies the following fundamental rules introduced by W.R Hamilton [1]:

$ij = k = -ji$, $jk = i = -kj$, and $ki = j = -ik$. Hence,

$$i^2 = j^2 = k^2 = ijk = -1$$

Unit quaternion provides a convenient mathematical notation for representing orientations in 3D space, so it's used to derive a quadrotor dynamic model. 3D Rotation matrix $R_I^B(q)$ that project BF quantities into IF can be obtained as [23]

$$\begin{bmatrix} x_i \\ y_i \\ z_i \end{bmatrix} = \begin{bmatrix} 2(q_o{}^2 + q_1{}^2) - 1 & 2(q_1 q_2 + q_o q_3) & 2(q_1 q_3 - q_o q_2) \\ 2(q_1 q_2 - q_o q_3) & 2(q_o{}^2 + q_2{}^2) - 1 & 2(q_2 q_3 + q_o q_1) \\ 2(q_o q_2 + q_1 q_3) & 2(q_2 q_3 - q_o q_1) & 2(q_o{}^2 + q_3{}^2) - 1 \end{bmatrix} \begin{bmatrix} x_b \\ y_b \\ z_b \end{bmatrix} \qquad (1)$$

The rate transformation matrix that projects BF rates into IF: $\dot{q}$ is obtained as

$$\begin{bmatrix} \dot{q}_o \\ \dot{q}_1 \\ \dot{q}_2 \\ \dot{q}_3 \end{bmatrix} = \frac{1}{2} \begin{bmatrix} q_o & -q_1 & -q_2 & -q_3 \\ q_1 & q_o & -q_3 & q_2 \\ q_2 & q_3 & q_o & -q_1 \\ q_3 & -q_2 & q_1 & q_o \end{bmatrix} \begin{bmatrix} 0 \\ p \\ q \\ r \end{bmatrix} \qquad (2)$$

2.2   Model Derivation

By using Newton's second law of rotational motion for fully actuated quadrotor rotational dynamics, the net moment derived from BF is as follows:

$$\sum \tau_i = J\alpha = [U_2, U_3, U_4]^T - M_{gyro} - M_{aero} \qquad (3)$$

$$J\dot{\omega} + \omega \times (J\omega) = [U_2, U_3, U_4]^T - M_{gyro} - M_{aero}$$

$$\dot{\omega} = J^{-1}\left[[U_2, U_3, U_4]^T - \omega \times (J\omega) - M_{gyro} - M_{aero} \right]$$

where: J = inertia matrix. $\dot{\omega} = [\dot{p}, \dot{q}, \dot{r}]^T$ are angular accelerations; U_2, U_3, and U_4 are moments which are used to control roll, pitch, and yaw, respectively; M_{gyro} = gyroscopic moment, and M_{aero} = aerodynamic frictional moment.

$$\dot{p} = \frac{1}{J_{xx}} U_2 + \frac{J_{yy} - J_{zz}}{J_{xx}} qr - \frac{J_p q}{J_{xx}} \Omega_r - \frac{C_{a_p}}{J_{xx}} p \qquad (4)$$

$$\dot{q} = \frac{1}{J_{yy}} U_3 + \frac{J_{zz} - J_{xx}}{J_{yy}} pr + \frac{J_p p}{J_{yy}} \Omega_r - \frac{C_{a_q}}{J_{yy}} q \qquad (5)$$

$$\dot{r} = \frac{1}{J_{zz}} U_4 + \frac{J_{xx} - J_{yy}}{J_{zz}} pq - \frac{C_{a_r}}{J_{zz}} rw \qquad (6)$$

By applying the Newton-Quaternion approach, translational dynamics is derived as:

$$\sum F_i = ma = R_I^B(q)U_1 - F_g - F_d$$

$$a = \frac{1}{m}[R_b{}^e U_1 - F_g - F_d]$$

$$\begin{bmatrix} \ddot{x} \\ \ddot{y} \\ \ddot{z} \end{bmatrix} = \frac{1}{m}\left(\begin{bmatrix} 2(q_o{}^2 + q_1{}^2) - 1 & 2(q_1 q_2 + q_o q_3) & 2(q_1 q_3 - q_o q_2) \\ 2(q_1 q_2 - q_o q_3) & 2(q_o{}^2 + q_2{}^2) - 1 & 2(q_2 q_3 + q_o q_1) \\ 2(q_o q_2 + q_1 q_3) & 2(q_2 q_3 - q_o q_1) & 2(q_o{}^2 + q_3{}^2) - 1 \end{bmatrix} \begin{bmatrix} 0 \\ 0 \\ U_1 \end{bmatrix} - \begin{bmatrix} 0 \\ 0 \\ mg \end{bmatrix} - \begin{bmatrix} C_{d_x} \dot{x} \\ C_{d_y} \dot{y} \\ C_{d_z} \dot{z} \end{bmatrix} \right)$$

$$\ddot{x} = 2(q_1q_3 - q_0q_2)\frac{U_1}{m} - \frac{C_{d_x}}{m}\dot{x} \qquad (7)$$

$$\ddot{y} = 2(q_2q_3 + q_0q_1)\frac{U_1}{m} - \frac{C_{d_y}}{m}\dot{y} \qquad (8)$$

$$\ddot{z} = \left[2({q_0}^2 + {q_3}^2) - 1\right]\frac{U_1}{m} - g - \frac{C_{d_z}}{m}\dot{z} \qquad (9)$$

where $a = [\ddot{x}, \ddot{y}, \ddot{z}]'$ are linear accelerations, $C_{d_x}, C_{d_y}, C_{d_z}$ are aerodynamic drag forces along x, y, z axis, U_1 is altitude control effort, m is total mass of quadrotor, and g is acceleration due to gravity.

Aerodynamic Ground Effect. Ground Effect (GE) is an aerodynamic phenomenon in which the ground pushes the quadrotor up when the propeller approaches the ground surface. For the same transmitted power to the actuator, the propeller generates more trust caused only by the presence of the ground [22] as shown in Fig. 2.

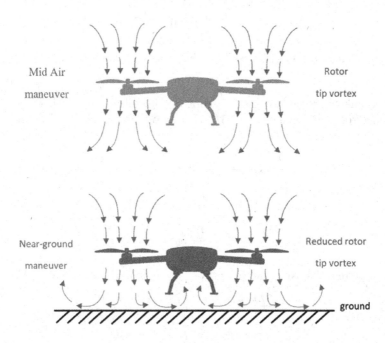

Fig. 2. Generated thrust Variation at Mid Air and Near-ground maneuvering [11]

Since the quadrotor has a stronger GE up to $z = 5R$, GE is measured for altitude up to $z = 6R$. So, the region where $z \geq 0.5R$ and $z \leq 6R$ is considered as In Ground Effect (IGE), and the region where $z < 0.5R$ and $z > 6R$ is considered as Out of Ground Effect (OGE) where GE is typically doesn't exist. Experiments show that at constant power, GE is expressed as [2,7]:

$$f_{GE}(z) = \begin{cases} \frac{1}{1-\gamma(\frac{R}{4z})^2} & 0.5R \leq z \leq 6R \\ 1 & otherwise \end{cases} \tag{10}$$

where $f_{GE}(z)$ is the ground effect factor, γ is the correction coefficient that is estimated from the experiment and is used to adjust unpredictable airflow influence between the rotors of the quadrotor, and R is the radius of the propellers. To account for GE in MATLAB simulation, the thrust force model is modified as [2]:

$$T_i = b\Omega^2 f_{GE}(z) \tag{11}$$

By combining Eqs. (4), (7) and (6) and Eqs. (7), (8) and (9), the translational and rotational flight dynamic equations of quadrotor based on the Newton-Quaternion approach in the presence of GE are summarized as:

$$\ddot{x} = 2(q_1q_3 - q_0q_2)\frac{U_1}{m}f_{GE} - \frac{C_{d_x}}{m}\dot{x} \tag{12}$$

$$\ddot{y} = 2(q_2q_3 + q_0q_1)\frac{U_1}{m}f_{GE} - \frac{C_{d_y}}{m}\dot{y} \tag{13}$$

$$\ddot{z} = [2(q_0{}^2 + q_3{}^2) - 1]\frac{U_1}{m}f_{GE} - g - \frac{C_{d_z}}{m}\dot{z} \tag{14}$$

$$\dot{p} = \frac{1}{J_{xx}}U_2 f_{GE}(z) + \frac{J_{yy} - J_{zz}}{J_{xx}}qr - \frac{J_{pq}}{J_{xx}}\Omega_r - \frac{C_{a_p}}{J_{xx}}p \tag{15}$$

$$\dot{q} = \frac{1}{J_{yy}}U_3 f_{GE}(z) + \frac{J_{zz} - J_{xx}}{J_{yy}}pr + \frac{J_p p}{J_{yy}}\Omega_r - \frac{C_{a_q}}{J_{yy}}q \tag{16}$$

$$\dot{r} = \frac{1}{J_{zz}}U_4 f_{GE}(z) + \frac{J_{xx} - J_{yy}}{J_{zz}}pq - \frac{C_{a_r}}{J_{zz}}r \tag{17}$$

$$\dot{q}_o = -\frac{1}{2}(pq_1 + qq_2 + rq_3) \tag{18}$$

$$\dot{q}_1 = \frac{1}{2}(pq_o + rq_2 - qq_3) \tag{19}$$

$$\dot{q}_2 = \frac{1}{2}(qq_o + pq_3 - rq_1) \tag{20}$$

$$\dot{q}_3 = \frac{1}{2}(rq_o + qq_1 - pq_2) \tag{21}$$

Based on Fig. 1, control mixing(or control allocation) equation is obtained as

$$\Omega_1 = \sqrt{\frac{1}{f_{GE}(z)}\left(\frac{U_1}{4b} + \frac{U_2}{4b\ell} + \frac{U_3}{4b\ell} - \frac{U_4}{4d}\right)} \tag{22}$$

$$\Omega_2 = \sqrt{\frac{1}{f_{GE}(z)}\left(\frac{U_1}{4b} - \frac{U_2}{4b\ell} + \frac{U_3}{4b\ell} + \frac{U_4}{4d}\right)} \tag{23}$$

$$\Omega_3 = \sqrt{\frac{1}{f_{GE}(z)}\left(\frac{U_1}{4b} - \frac{U_2}{4b\ell} - \frac{U_3}{4b\ell} - \frac{U_4}{4d}\right)} \tag{24}$$

$$\Omega_4 = \sqrt{\frac{1}{f_{GE}(z)}\left(\frac{U_1}{4b} + \frac{U_2}{4b\ell} - \frac{U_3}{4b\ell} + \frac{U_4}{4d}\right)} \tag{25}$$

and the overall residual or net propeller angular speed Ω_r is obtained as

$$\Omega_r = -\Omega_1 + \Omega_2 - \Omega_3 + \Omega_4 \tag{26}$$

3   Intelligent Flight Control Design of Quadrotor UAV

Flight Control System (FCS) is the brain of the quadrotor that performs the tasks of pilot. It is responsible for achieving the desired attitude and position while the quadrotor flies. Quadrotor control researchers face many challenges in developing high-performance controllers due to time-varying properties of the environment, nonlinear coupled dynamics of the system, uncertainties in modeling, and external disturbances from the wind. These make the dynamics more challenging for control applications. To design a high-performance controller under those circumstances, in this study, an intelligent flight controller is designed by combining adaptive control technique with the learning power of ANN; which is two equivalent topics under different umbrellas of research in the Control Engineering arena: control systems and machine learning, respectively.

ANNs for Identification and Control. ANNs with a sufficient number of network parameters: several hidden layers and neurons in those hidden layers, have well-proved properties to universally approximate any smooth arbitrary nonlinear functions with the desired level of accuracy. This is the core property of ANNs for their superiority over other function approximation mechanisms. Furthermore, once trained, ANNs require less computation time and memory storage to perform the intended tasks [24]. However, there is one critical issue, that is, choosing the right number of network parameters since there is no defined rule to select them. Therefore, most of the time they are chosen by trial and error by keeping in mind that, the smaller the numbers are, the better the network is in terms of memory storage and processing time.

ANN has the advantage that it doesn't require mathematical equations for its learning. In fact, it requires input and output data only. Based on this data, ANN through training, learns and understands the relationship between input and output data. These generalization and learning capabilities of ANN, make it to be used in many applications as a Black-box identifier and controller. Dynamic systems can be better approximated by including past input and past output with the present inputs since additional information is contained in them. Such a model of the system is considered Autoregressive since its output depends on the current and past values of its input and outputs.

Feed-Forward Neural Network (FFNN) does not have a memory to memorize its previous state, and it only considers the current input, so it can not handle sequential data. For sequential data, Recurrent Neural Networks(RNNs) are the best candidate since they have a memory that helps them to store previous inputs and output data to generate the next output of the sequence. Among different classifications of RNN, Hopfield RNNs have been proven to be more effective at handling complexity and nonlinearity issues. From Hopfield RNNs, a Nonlinear Autoregressive Network with Exogenous Inputs (NARX) is commonly used in time-serious modeling. In this modeling, the current output depends on the past output and input with the inclusion of the current input.

Quadrotor Controller Design Strategy. Due to the underactuation of quadrotor translational dynamics, there is no way to directly control the lateral motion of the quadrotor. So, a nested control approach is implemented in which desired attitude commands are generated by the outer position control loop. Then, the inner attitude control loop stabilizes the quadrotor based on the generated desired command, accordingly. To ensure that every quadrotor output follows a specified trajectory, virtual control inputs are introduced to the system. The physical interpretation of this control law is that the control of the translation motion of the quadrotor depends on desired quaternion rotation q_d, and total thrust force U_1. In the case of no external disturbance acts on the system, these virtual control inputs are derived as follows:

$$\begin{bmatrix} U_x \\ U_y \\ U_z \end{bmatrix} = R_I^B(q_d) \begin{bmatrix} 0 \\ 0 \\ U_1 \end{bmatrix} f_{GE} - \begin{bmatrix} 0 \\ 0 \\ mg \end{bmatrix} \tag{27}$$

$$U_x = 2(q_{1_d}q_{3_d} - q_{0_d}q_{2_d})U_1 f_{GE} \tag{28}$$

$$U_y = 2(q_{2_d}q_{3_d} + q_{0_d}q_{1_d})U_1 f_{GE} \tag{29}$$

$$U_z = \left[2(q_{0_d}{}^2 + q_{3_d}{}^2) - 1\right] U_1 f_{GE} - mg \tag{30}$$

where U_x, U_y and U_z are virtual position control inputs that control quadrotor x, y, and z maneuvering. This method of controlling a nonlinear system as linear is called Nonlinear Dynamic Inversion(NDI), which is a very popular method employed in aircraft automatic FCS's. Hence, by substituting Eqs. (28), (29) and (30) into Eqs. (12), (13) and (14) in terms of desired quaternion, translational quadrotor dynamics can be rewritten in terms of virtual control inputs as

$$\ddot{x} = \frac{U_x}{m} - \frac{C_{d_x}}{m}\dot{x} \tag{31}$$

$$\ddot{y} = \frac{U_y}{m} - \frac{C_{d_y}}{m}\dot{y} \tag{32}$$

$$\ddot{z} = \frac{U_z}{m} - \frac{C_{d_z}}{m}\dot{z} \tag{33}$$

By assigning those virtual control inputs, quadrotor dynamics become fully actuated I.e., each trajectory of the quadrotor is controlled separately by their corresponding control input. Based on this setup, the novel control architecture that this research study contributed is shown in Fig. 3.

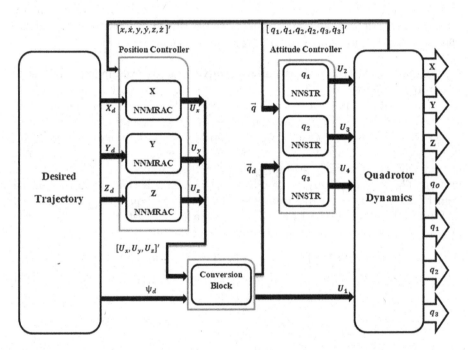

Fig. 3. Hierarchical NNMRAC and NNSTR hybrid control architecture of Quadrotor

Based on Eq. (31), the conventional Model Reference Adaptive Control (MRAC) is designed, first to control the position of the quadrotor [4]. Then, by exporting input-output data of adaptation law to MATLAB® workspace, FFNN is trained to estimate controller parameters online as a control architecture in Fig. 4.

Desired U_1 and q_d, which are manipulated inside the conversion block, are derived by rearranging Eqs. (28), (29) and (30). Finally, they can be obtained

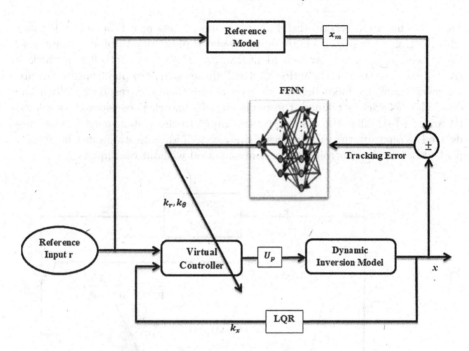

Fig. 4. ANN-based MRAC of quadrotor position control architecture

as:

$$U_1 = \frac{1}{f_{GE}}\sqrt{U_x^2 + U_y^2 + (U_z + mg)^2} \tag{34}$$

$$q_{o_d} = \frac{1}{\sqrt{2}}\sqrt{\frac{U_z + mg}{\sqrt{U_x^2 + U_y^2 + (U_z + mg)^2}} - 2sin^2(\frac{\psi_d}{2}) + 1} \tag{35}$$

$$q_{1_d} = \frac{sin(\frac{\psi_d}{2})U_x + \frac{1}{\sqrt{2}}U_y\sqrt{\frac{U_z+mg}{\sqrt{U_x^2+U_y^2+(U_z+mg)^2}} - 2sin^2(\frac{\psi_d}{2}) + 1}}{U_z + mg + \sqrt{U_x^2 + U_y^2 + (U_z + mg)^2}} \tag{36}$$

$$q_{2_d} = \frac{sin(\frac{\psi_d}{2})U_y - \frac{1}{\sqrt{2}}U_x\sqrt{\frac{U_z+mg}{\sqrt{U_x^2+U_y^2+(U_z+mg)^2}} - 2sin^2(\frac{\psi_d}{2}) + 1}}{U_z + mg + \sqrt{U_x^2 + U_y^2 + (U_z + mg)^2}} \tag{37}$$

$$q_{3_d} = sin(\frac{\psi_d}{2}) \tag{38}$$

On the other hand, the attitude controller has a one-to-one relationship between control inputs and their corresponding outputs i.e., there is direct actuation control for the attitude of the quadrotor that drives the quadrotor in the desired direction. Hence, the RNN-based indirect adaptive controller (RNNSTR) is designed to control the attitude of the quadrotor as a control architecture in

Fig. 5. In this architecture, the RNN plant model acts as a Black-box identifier which is trained first to generate one step ahead prediction of plant output, and the process is known as system identification. Then, the controller network is trained to generate control action that will drive estimated plant output toward reference input by following the desired reference model response. When the controller network is trained, the plant model network is considered as a fixed network. To stabilize the system, control input to the system must be one step delayed to separate the two processes since system identification must be accomplished before the controller takes control action without overlapping.

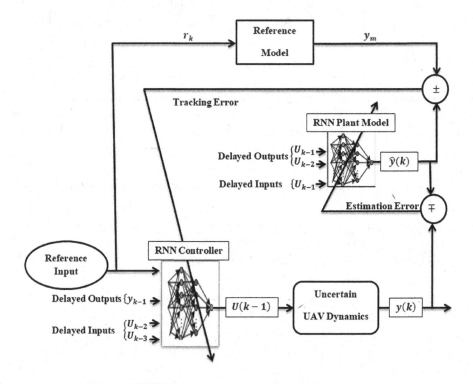

Fig. 5. RNNSTR quadrotor attitude control architecture

Quadrotor Position Controller Design. Plant dynamics with matched uncertainty is given as

$$\dot{x} = Ax + B(U - \theta^{*T}\phi(x)) \tag{39}$$

where $f(x) = \theta^{*T}\phi(x)$ is matched uncertainty, $\theta = [\theta_1, \theta_2...]$ are unknown constant vector, and $[\phi_1(x), \phi_2(x)...]$ are vector of known bounded basis functions. And reference model for 2nd order system is given as:

$$\dot{x}_m = A_m x_m + B_m r \tag{40}$$

Ideal control law when $x(t)$ is perfectly track $x_m(t)$, is defined as

$$U^* = k_r^* r - k_x^* x + \theta^{*T} \phi(x) \tag{41}$$

An actual control law that perfectly cancels out the matched uncertainty is defined as

$$U = k_r r(t) - k_x x(t) + \theta^T \phi(x) \tag{42}$$

where k_r, k_x, and θ are adaptation parameters; and estimation errors are defined as $\tilde{k}_x = k_x - k_x^*$, $\tilde{k}_r = k_r - k_r^*$ and $\tilde{\theta} = \theta - \theta^*$. To drive adaptation law by using the Lyapunov direct method, let's choose a positive definite Lyapunov candidate function of tracking and estimation error from [14] as follows:

$$V(e, \tilde{k}_r, \tilde{k}_x, \tilde{\theta}) = e^T P e + |b| \left(\frac{\tilde{k}_r^{\ 2}}{\gamma_r} + \tilde{k}_x \gamma_x^{-1} \tilde{k}_x^{\ T} + \tilde{\theta}^T \gamma_\theta^{-1} \tilde{\theta} \right) \tag{43}$$

For closed-loop system to be globally stable, the gradient of the Lyapunov function of Eq. (43) must be negative. Hence, adaptation laws are derived in the way that $\dot{V} < 0$ as:

$$\dot{k}_r = \gamma_r r e^T \bar{P} sgn(b) \tag{44}$$

$$\dot{k}_x = -\gamma_x x e^T \bar{P} sgn(b) \tag{45}$$

$$\dot{\theta} = \gamma_\theta \phi e^T \bar{P} sgn(b) \tag{46}$$

Reference model for 2nd order system is designed based on desired transient response with settling time of 3 second and 5% maximum overshoot:

$$x_m = \frac{4}{s^2 + 2.76\,s + 4} r \tag{47}$$

Since the control problem is a tracking problem, the error dynamics of translational motion along the x-axis in Eq. (31) to design LQR becomes

$$\dot{z}_1 = z_2 \tag{48}$$

$$\dot{z}_2 = -\frac{C_{d_x}}{m} z_2 - \frac{U_x}{m} \tag{49}$$

$$U_p = U_x = U_y = U_z = U_{LQR} + U_{adaptive}$$

Quadrotor Attitude Controller Design. To design a controller, a small quaternion approach is implemented to represent rotational dynamics in terms of quaternion; to decouple the controlled variables, Coriolis term and gyroscopic effects are ignored. Furthermore, quadrotor navigation altitude is in the region of OGE, so $f_{GE} = 1$. Therefore, Eqs. (12), (13), (14), (15) and (21) are reduced into

$$\ddot{q}_1 = \frac{1}{J_{xx}} \left(\frac{1}{2} U_2 - C_{a_p} \dot{q}_1 \right) \tag{50}$$

$$\ddot{q}_2 = \frac{1}{J_{yy}} \left(\frac{1}{2} U_3 - C_{a_q} \dot{q}_2 \right) \tag{51}$$

$$\ddot{q}_3 = \frac{1}{J_{zz}} \left(\frac{1}{2} U_4 - C_{a_r} \dot{q}_3 \right) \tag{52}$$

From the property of unit quaternion, the actual scalar part of quaternion q_o will be converged to its desired q_{o_d} by the following relation:

$$q_o = \sqrt{1 - (q_1{}^2 + q_2{}^2 + q_3{}^2)} \tag{53}$$

Based on Fig. 5 to train the plant model network, the maximum-minimum input of the system is obtained first by considering 75% of the maximum speed of the BLDC motor of 10000 RPM, and this results 1.1/−1.1 Nm and desired output in the range of [−1 1]. Then, 5000 samples of input-output data are generated with a sampling time of 0.05 s. Finally, a three-layer ANN with 10 neurons in the hidden layer is trained and analyzed as shown in Figs. 6, 7 and 8.

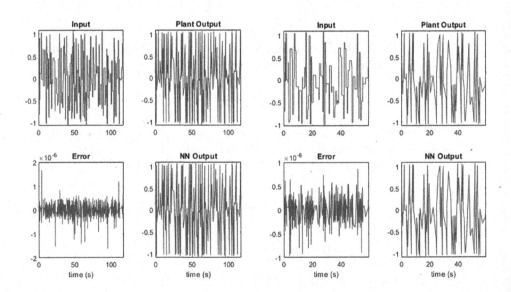

Fig. 6. Training   **Fig. 7.** Validation

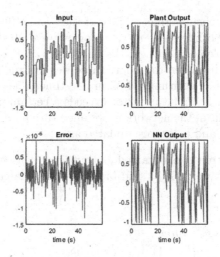

Fig. 8. Testing

To train the controller network, 4000 data are generated. The desired response and tracking performance of the attitude controller for uniform random reference input signal on the training dataset is shown in Fig. 9. The result shows that the plant network output(green) perfectly tracks the desired reference model output(blue). By applying the same procedure, RNNSTR attitude controllers for q_2 and q_3 are designed by using the same procedures.

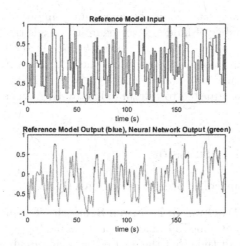

Fig. 9. Tracking performance of RNNSTR on generated dataset

Online Learning Algorithm. It's used to update network parameters in real-time. Using initial sets of network parameters from offline training increases the convergence rate and ensures the stability of the system [9]. Hence, to increase ANN's performance, both learning methodologies are applied in this study. Here, offline training is implemented to get the initial sets of weights and biases, and online training is implemented to update those weights and biases by a generalized Back-propagation Algorithm (BPA) in real-time. These weights and biases are updated only when the output error exceeds a certain predefined threshold value. Once the output of the network is obtained through forward propagation, output layer weights, and biases are updated as

- Once the ANN output y_i is obtained, output layer delta δ_j is computed, first. Then, output layer weights w_{ji} and biases b_j are updated as follows:

$$\delta_j = y_i(1 - y_i)e_o \tag{54}$$
$$w_{ji}(k + 1) = w_{ji}(k) + \mu\delta_j(k)x_i(k) \tag{55}$$
$$b_j(k + 1) = b_j(k) + \mu\delta_j(k) \tag{56}$$

where delta δ is the rate at which the error changes w.r.t the output of a particular neuron in ANN.

- In BPA, the hidden layer error can be obtained as follows:

$$e_h = \sum_{j=1}^{N_o}\sum_{h=1}^{M_o} w_{kh}\delta_j \tag{57}$$

where μ is the learning rate, N_o is the maximum number of output nodes, M_o is the maximum number of hidden layer neurons, w_{kh} is the weights of hidden to the output layer, and e_h is the error at each node in the hidden layer. Once hidden layer error and delta are computed, then the delta rule in Eqs. (54) to (56) is applied similarly to update the hidden layer weights and biases. In this study, the network is adjusted to retrain online by BPA when the error exceeds ± 0.0005.

4   Results and Discussion

4.1   Designed Controller Performance Analysis

Main Trajectory Tracking in Nominal Scenario. The main trajectory generation algorithm is developed specifically for aerial photography for image acquisition by setting the altitude that the quadrotor navigates above the crop at 9m, navigation speed as $2\,\text{m/s}$, and lateral overlap i.e., the area of one image includes the area already captured in another image, as 20%.

As can be seen from the Figs. 10a and 10b, the 3D main trajectory and rotational quadrotor trajectories precisely track the desired reference model response; 3D trajectory tracking in Fig. 10 also proves that the controller guarantees accurate desired reference model response tracking by perfectly stabilizing the quadrotor when it flies on the top of the field as a desired stated path. Furthermore, the control inputs results in Fig. 11 show that the quadrotor for the main trajectory is controlled near to hovering position i.e., $U_1 \approx 4.591N = mg$, and the control effort required to drive the quadrotor is smooth and minimal that prove feasibility of the proposed control algorithm. One thing we can notice here is that the more the turning point that the trajectory has, the more control effort is required. Therefore, to minimize control effort, desired trajectory pose is better to be designed with fewer turning points.

Performance Analysis by Imposing Matched Uncertainties. Aerodynamic drag coefficients and inertia depend on the speed of the quadrotor, the density of air, and the altitude of flight, so it's crucial to design a controller that can tolerate these parametric uncertainties in the quadrotor model. Therefore, to validate the robustness of the proposed control algorithm in the presence of parametric variations that can be considered as a matched uncertainty, parametric variations w.r.t time are conducted as follows: $J_{xx} = 2J_{xx_o}$, $J_{yy} = 2J_{yy_o}$, $J_{zz} = 2J_{zz_o}$, $c_{d_x} = c_{d_y} = c_{d_z} = (1+0.1t)c_{d_o}$, and $c_{a_p} = c_{a_q} = c_{a_r} = (1+0.01t)c_{a_o}$ where subscript 'o' implies parameters at the nominal scenario. The proposed controller tracking performance in this scenario is shown in Fig. 12.

Performance Analysis by Injecting Unmatched Uncertainty. Since quadrotor maneuvering in the field is exposed to external disturbance from the wind, performance analysis is done by applying unknown external disturbances as unmatched uncertainties as $d = \sin(0.03t)$, in the direction of the most critical translation path. Comparison is done in Fig. 13 based on the tracking performance of online-tuned ANN-based MRAC, offline-tuned ANN-based MRAC, and conventional MRAC. Notice that disturbances that we consider here are arbitrary functions since the uncertainty is assumed as unmatched, the controller doesn't have knowledge about it. As can be seen from Fig. 13, for slowly varying disturbance the proposed controller with online tuned Neural Network parameters, which update its parameters online till the output error is within predefined threshold value i.e., $|e| \leq 0.0005$, shows the promising result by rejecting slowly varying applied external disturbance when compared to the others presented controllers. Therefore, the developed online learning algorithm enhances the external disturbance rejection capability of the designed control system.

4.2   CNN Training for Yellow Wheat Rust Detection

Given the highly destructive nature of yellow wheat rust in crops, various researchers employ different methods to detect this wheat disease early and minimize its impact. Among those [15,20] proposed a deep learning approach for yellow wheat rust recognition. In [15], a team of researchers from Turkey in

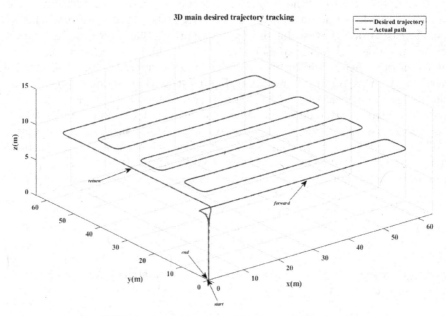

(a) 3D main trajectory tracking in nominal scenario

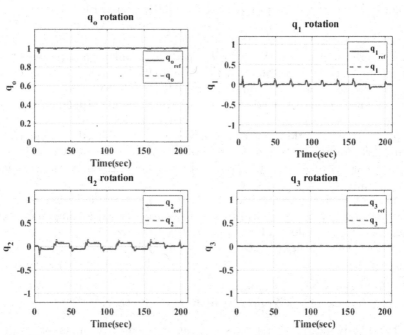

(b) Quaternion-based rotational trajectory tracking in nominal scenario

Fig. 10. Trajectory tracking performance in nominal scenario

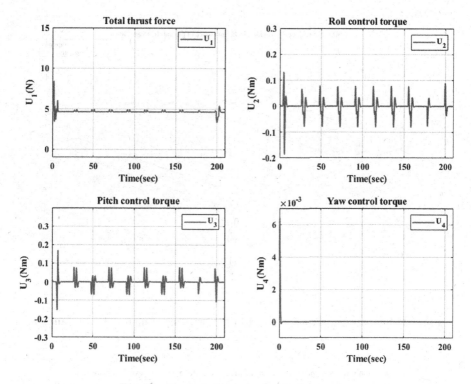

Fig. 11. Control efforts in nominal scenario

collaboration with experts created a database of yellow wheat rust with different severity levels. The dataset consists of raw and pre-processed images with categories of healthy and yellow wheat rust-diseased leaf images. In this paper, state-of-the-art CNN is implemented for image analysis and achieved 91% accuracy. In this study, the image analysis part is the extension of their work, but here, the whole dataset is used to train the deep learning model to increase model accuracy since model accuracy is directly related to the amount of data that is used to train the network. The yellow wheat rust image dataset contains six classes of the severity level of images: No disease, Resistant, Moderately Resistant, Moderately Resistance Moderately Susceptible, Moderately Susceptible, and Susceptible [15]. By using the transfer learning technique, the retrained Xception CNN model for image analysis yields a training accuracy of 97.28%, as the performance illustrated in Figs. 14a and 14b.

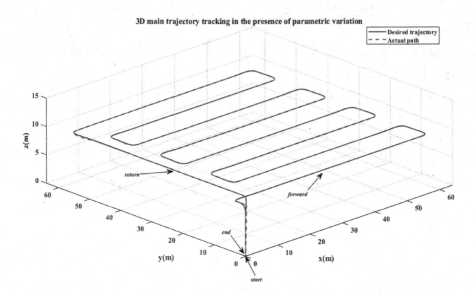

Fig. 12. 3D main trajectory tracking in the presence of matched uncertainty

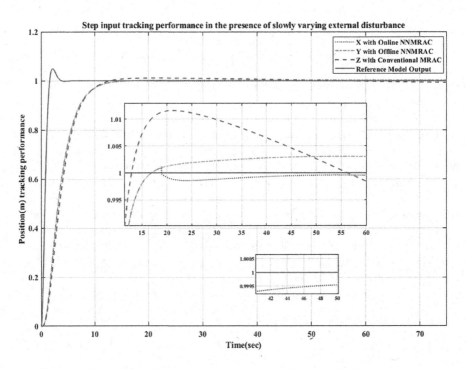

Fig. 13. Tracking performance in the presence of unmatched uncertainty

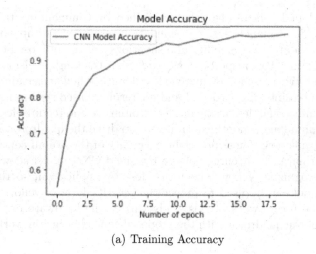

(a) Training Accuracy

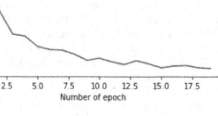

(b) Cross-Entropy loss function

Fig. 14. Performance of the retrained Xception CNN model

5 Conclusion

In this paper, the following main tasks have been performed: modeling quadrotor dynamics, designing conventional MRAC for position control, training FFNN to estimate controller parameters, training RNN for system identification and attitude control, developing online learning algorithm to enhance performance of controller, verifying robustness of proposed controller approach by imposing matched and unmatched uncertainties, training CNN by using transfer learning approaches, and comparative analysis based on MATLAB® simulation results between presented control techniques.

The developed control algorithm is verified by commanding the system to track various flight maneuvers. The simulation results, w.r.t implemented flight maneuvers, validate the feasibility and superiority of the proposed control technique in achieving fast response speed and high tracking precision in nominal scenarios and the presence of matched and unmatched uncertainties. Due to the implemented intelligent control and quaternion-based system modeling, the overall system becomes less mathematically intensive and requires less processing time and minimal and smooth control effort to drive the quadrotor according to the desired trajectory. These prove the feasibility of the overall control architecture. Furthermore, the accuracy of the retrained CNN model shows promising results in recognizing yellow wheat rust for PA application, so the retrained Xception model can be used to detect the severity level of yellow wheat rust. Therefore, the future work of this study will be the hardware implementation and testing of the quadrotor with the proposed controller in real-world scenarios.

References

1. Abaunza Gonzalez, H.: Robust tracking of dynamic targets with aerial vehicles using quaternion-based techniques. Ph.D. thesis, Compiègne (2019)
2. Ahmadinejad, F., Bahrami, J., Menhaj, M.B., Ghidary, S.S.: Autonomous Flight of Quadcopters in the Presence of Ground Effect. In: 2018 4th Iranian Conference on Signal Processing and Intelligent Systems (ICSPIS), pp. 217–223. IEEE (2018)
3. Alemaw, G., Agegnehu, G.: Precision agriculture and the need to introduce in Ethiopia. Ethiopian J. Agric. Sci. **29**(3), 139–158 (2019)
4. Assefa, A.: Neural network based direct MRAC technique for improving tracking performance for nonlinear pendulum system. J. Inf. Electr. Electron. Eng. (JIEEE) 1, November 2020. https://doi.org/10.54060/JIEEE/001.02.004
5. Bakshi, N.A.: Model reference adaptive control of quadrotor UAVs: a neural network perspective. Adaptive Robust Control Systems, p. 135 (2018)
6. Benosman, M.: Model-based vs data-driven adaptive control: an overview. Int. J. Adapt. Control Sig. Process. **32**(5), 753–776 (2018)
7. Bernard, D.D.C., Riccardi, F., Giurato, M., Lovera, M.: A dynamic analysis of ground effect for a quadrotor platform. IFAC-PapersOnLine **50**(1), 10311–10316 (2017)
8. Daponte, P., et al.: A review on the use of drones for precision agriculture. In: IOP Conference Series: Earth and Environmental Science, vol. 275, p. 012022. IOP Publishing (2019)
9. De mel, W.R.: Artificial neural network based adaptive controller for DC motors (2004)
10. Elias, E., et al.: Cereal yields in Ethiopia relate to soil properties and n and p fertilizers. Nutr. Cycl. Agroecosyst. **126**, 1–14 (2023)
11. Emran, B.J., Najjaran, H.: A review of quadrotor: an underactuated mechanical system. Annu. Rev. Control. **46**, 165–180 (2018)
12. Fresk, E.: Modeling, control and experimentation of a variable pitch quadrotor (2013)
13. Fresk, E., Nikolakopoulos, G.: Full quaternion based attitude control for a quadrotor. In: 2013 European Control Conference (ECC), pp. 3864–3869. IEEE (2013)

14. Gupta, D., Kumar, A., Giri, V.K.: Effect of adaptation gain and reference model in MIT and lyapunov rule-based model reference adaptive control for first-and second-order systems. Trans. Inst. Measure. Control 01423312231203483 (2023)
15. Hayit, T., Erbay, H., Varçın, F., Hayit, F., Akci, N.: Determination of the severity level of yellow rust disease in wheat by using convolutional neural networks. J. Plant Pathol. **103**(3), 923–934 (2021)
16. Lv, Y.Y., Huang, W., Liu, J., Peng, Z.F.: A sliding mode controller of quadrotor based on unit quaternion. In: Applied Mechanics and Materials, vol. 536, pp. 1087–1092. Trans Tech Publ (2014)
17. Maddikunta, P.K.R., et al.: Unmanned aerial vehicles in smart agriculture: applications, requirements, and challenges. IEEE Sens. J. **21**(16), 17608–17619 (2021)
18. Mogili, U.M.R., Deepak, B.B.V.L.: Review on application of drone systems in precision agriculture. Procedia Comput. Sci. **133**, 502–509 (2018)
19. Mohidem, N.A., et al.: Application of multispectral UAV for paddy growth monitoring in Jitra, Kedah, Malaysia. In: IOP Conference Series: Earth and Environmental Science, vol. 1038, p. 012053. IOP Publishing (2022)
20. Pan, Q., Gao, M., Wu, P., Yan, J., Li, S.: A deep-learning-based approach for wheat yellow rust disease recognition from unmanned aerial vehicle images. Sensors **21**(19), 6540 (2021)
21. Rzevski, G.: Artificial intelligence in engineering: past, present and future. WIT Trans. Inf. Commun. Technol. **10**, 14 (1970)
22. Sanchez-Cuevas, P., Heredia, G., Ollero, A.: Characterization of the aerodynamic ground effect and its influence in multirotor control. Int. J. Aerospace Eng. **2017** (2017)
23. Sanwale, J., Trivedi, P., Kothari, M., Malagaudanavar, A.: Quaternion-based position control of a quadrotor unmanned aerial vehicle using robust nonlinear third-order sliding mode control with disturbance cancellation. Proc. Inst. Mech. Eng. Part G: J. Aerosp. Eng. **234**(4), 997–1013 (2020)
24. Shin, Y.: Neural network based adaptive control for nonlinear dynamic regimes. Ph.D. thesis, Georgia Institute of Technology (2005)

AI Ethics and Life Sciences

Systems Thinking Application to Ethical and Privacy Considerations in AI-Enabled Syndromic Surveillance Systems: Requirements for Under-Resourced Countries in Southern Africa

Taurai T. Chikotie[1]($\boxtimes$)(ID), Bruce W. Watson[1](ID), and Liam R. Watson[2](ID)

[1] Centre for AI Research, School for Data-Science and Computational Thinking, Stellenbosch University, Cape Town, South Africa
`taurai.chikotie@icloud.com, bruce.watson@cair.org.za`
[2] Computer Science Department, University of Waterloo, Ontario, Canada

Abstract. This paper examines the ethical and privacy considerations of implementing AI-enabled syndromic surveillance systems in under-resourced Southern African countries. It highlights the rise of digital health technologies and big data in public health, which enables early disease outbreak detection and monitoring. In Southern Africa, where health systems are uniquely complex and dynamic, such advancements pose a chance to raise significant ethical and privacy concerns. The study employed a multifaceted methodology, combining literature review, systems dynamics modelling (SDM), systems thinking, and causal loop diagrams (CLDs) to analyse these issues. The research focuses on country-specific case studies to gain insights into the local context. Findings from the study identified factors such as data collection and privacy, data quality, algorithm performance and bias, regulatory frameworks, surveillance infrastructure, data sharing, community trust, capacity building, and education as critical when considering ethics and privacy issues in AI-enabled syndromic surveillance systems. In conclusion, the study emphasises the application of systems thinking principles and methodologies as a comprehensive approach to address these ethical and privacy challenges in Southern Africa. By considering the complex interplay of stakeholders, potential risks, and specific regional needs, the paper advocates for responsible and beneficial deployment of AI surveillance systems, ensuring the protection of individual rights and public interests.

Keywords: AI-enabled · Ethics and Privacy · Syndromic Surveillance

1 Introduction

With the global digital ecosystem on accelerated growth, emphasis has been put on the transformative potential of information and communications technology,

T. G. Debelee et al. (Eds.): PanAfriConAI 2023, CCIS 2069, pp. 197–218, 2024.
https://doi.org/10.1007/978-3-031-57639-3_9

coupled with global interconnectedness, to expedite human advancement, ameliorate the digital disparity, and cultivate knowledge-driven societies [1]. A proliferation of digital health technologies bringing with it fertile grounds for big data in public health has facilitated the development and deployment of electronic disease surveillance systems and as such, AI-enabled syndromic surveillance systems, which offer promising capabilities for early detection and monitoring of disease outbreaks not only in developed countries but also in under-resourced countries [2]. While the definition of an "under-resourced" country in the context of AI and disease surveillance is complex and multifaceted, this study assumes that it encompasses financial pressure, suboptimal healthcare service delivery, underdeveloped infrastructure, limited knowledge, research challenges, restricted social resources, geographical and environmental factors, human resource limitations, and the influence of beliefs and practices [65]. Such challenges have led to the under-reporting and under-ascertainment of infectious diseases further complicating the situation, and requiring careful selection of appropriate methods for estimating the true incidence of disease [66].

Based on the WHO's triple billion targets for 2019–2023, the disruptive nature of digital health technologies is identified as playing an important in benefiting the masses from Universal Health Coverage; health emergencies (including pandemics) protection, and preparedness; and improved health outcomes for a billion more people [3]. Forecasts as of 2021 indicate that the global digital market was worth around USD$195.1 Billion and it is expected to reach US$1.1 Trillion by the year 2032 at a compound annual growth rate (CAGR) of 22% [4].

Although still in its infancy stages in Southern Africa and with very limited statistics provided, the market is expected to grow as countries seek alternative ways of arresting the higher burden of disease in the region. South Africa is presumed to unlock an estimated US$11 billion in efficiency gains at an annual growth rate of 8.82% (CAGR 2023–2028) by the year 2028 in investment, while Zimbabwe will see a projected annual rate of 13.53% (CAGR 2023–2028) with an estimated market growth of US$179.7million in investment [5–7].

Investment towards strengthening disease surveillance and preparedness technologies is presumed to reach between US$285 billion to US$430 billion globally with estimated spending rates of 27% at a global and regional level and 73% at the country level (8% towards developed countries and 65% towards low-to-middle income countries (LMIC)) by the year 2030 [8]. It is, therefore, critical to acknowledge the transformative potential of Artificial Intelligence in the realm of healthcare, particularly in the context of disease surveillance systems. This recognition extends specifically to AI-enabled applications within syndromic disease surveillance, which represent a significant advancement in monitoring and managing public health [9,10]. These systems have played a crucial role in predictive analytics to understand the transmission and possibly curb the spread of infection, especially in a situation when health systems are strained [11].

As indicated in previous digital health investment reports, Southern African countries are projected to deploy more digital technologies in health however, the variations in the estimated investment figures are a testament to the unique and dynamic socio-technical environments between identified countries (South Africa and Zimbabwe). While the Southern African region has the highest internet penetration rates in Africa (72%), the two countries still have relative challenges in the high burden of disease, infrastructure support, skills shortages, funding challenges, and political leadership deficiencies amongst many other factors affecting mostly the respective public health systems [12–16]. Leveraging AI-enabled disease surveillance systems will help improve the efficiency of service delivery for these countries.

There is, however, a need to understand that while the blending of AI and healthcare innovation will help in improving decision-making and patient outcomes, the reality is much more refined, complex, and dynamic due to issues attached to such rapidly evolving technologies [17]. Such complexities advocate for the engagement of a hybrid of approaches to fully understand the various dynamics in the consideration of ethics and privacy in AI-enabled disease surveillance within an under-resourced context and systems thinking provides an avenue to understand and work with such complexities. An approach that has been applied in many healthcare-related studies including in AI, it helps provide insights into the relationships, interactions, feedback, and processes between variables constituting a particular system [18–21]. In the context of this paper, the complexity lies in understanding how the different variables under privacy and ethics in AI are interconnected and where possible clash while still affording room for a comprehensive model for responsible AI in disease surveillance within Southern Africa.

Experiences from developed countries suggest that the widespread implementation of AI-enabled disease surveillance systems has the potential to violate some of the important ethical and privacy considerations [11, 22]. This study in applying the systems thinking approach seeks to examine the ethical implications and privacy concerns associated with the implementation of AI-enabled syndromic surveillance systems in Southern Africa.

1.1 AI in Disease Surveillance

Globally, countries continue to be affected by emerging and re-emerging infectious diseases leading to unexpected high morbidity and mortality rates. As with the proliferation of technological innovations, the health fraternity has been leveraging precision tools and techniques to better understand and predict pandemics, and as such AI-enabled syndromic surveillance systems [10, 23, 24]. AI-enabled syndromic surveillance systems have been used in the developed world and such systems are all for early detection, prediction, and reporting of potential infectious disease cases thus helping with control of a pandemic. Notable AI applications that have contributed immensely to predicting and managing disease spread include, BlueDot - an application that helped in the prediction of the COVID-19 outbreak in 2019 and the Zika Virus; and ARGO (AutoRegression with GOogle search data) - leverages inter-net search data for the surveillance of flu-like illnesses providing valuable and faster insights than traditional disease surveillance methods [10, 25–28].

South Africa and Zimbabwe countries located in the Southern African region, face huge challenges of emerging and re-emerging diseases and ought to leverage technologies such as AI-enabled syndromic surveillance systems. In the last decade, the countries have had disease outbreaks such as cholera, typhoid, measles, mumps, and other respiratory diseases such as SARS [29–34]. Although acknowledging the positive contributions of AI-enabled syndromic surveillance systems, developing countries need to understand the application of these systems responsibly by addressing possible privacy and ethical challenges.

A comprehensive overview of the evidence on AI ethics and privacy concerns reveals a range of issues such as consent, algorithmic bias, transparency, respect for autonomy, beneficence, community trust, data protection, data anonymisation, data utilisation, data sharing, surveillance scope, and proportionality [10,11,35–38]. These concerns are particularly relevant in the context of AI applications, which have the potential to impact society significantly.

2   Methodology

Informed by insights from case studies in developed countries, this study adopts a multifaceted methodological approach, integrating a comprehensive literature review, systems dynamic modeling, and systems thinking.

2.1   Study Research Strategy

Literature Review. The literature review for this study was meticulously designed to encompass a broad spectrum of topics related to AI-enabled syndromic surveillance systems, with a focus on ethical and privacy considerations in the context of under-resourced Southern African countries. The study employed the following systematic approach to provide for a robust foundation while enabling an in-depth understanding of the multifaceted issues surrounding the implementation of AI-enabled syndromic surveillance systems:

Defining Key Themes: The review was structured around several core themes derived from the research objectives: digital health technologies, AI in public health, ethical considerations in AI, privacy concerns in digital surveillance, and the specific challenges faced by health systems in Southern Africa.

Search Strategy: A comprehensive search was conducted across multiple databases, including PubMed, Scopus, IEEE Xplore, and Google Scholar. Key search terms included combinations of "AI-enabled syndromic surveillance," "digital health ethics," "privacy in health technology," "AI in low-resource settings," and "healthcare system challenges in Southern Africa."

Inclusion and Exclusion Criteria: Studies were selected based on relevance to the themes, with a focus on recent reports, publications, and policy documents that provide insights into the latest developments in the field. Both empirical and theoretical works were included to ensure a holistic understanding. Exclusion criteria included non-peer-reviewed articles, outdated studies, and research not pertinent to the Southern African context.

Synthesis of Literature: The literature was synthesised to identify common findings, contrasting views, and emerging trends. This synthesis informed the development of the systems dynamics model and the construction of causal loop diagrams, ensuring that the study was grounded in comprehensive, current, and relevant scholarly work.

Country-Specific Case Studies: To tailor the review to the specific context of Southern Africa, country-specific case studies were examined. This included an analysis of local health system reports, policy documents, and studies conducted in the region, providing insights into the unique challenges and needs of these countries.

Continuous Review Process: The literature review was an ongoing iterative process, revisited throughout the study to incorporate new findings and perspectives that emerged during the research.

Building on this, the study applied systems thinking to dissect and understand the complex, adaptive nature of these surveillance systems. Systems thinking allowed for exploring the intricate interconnections and feedback loops inherent in the ethical and privacy dimensions of AI-enabled surveillance. This approach was pivotal in recognising the multifaceted and dynamic interactions within these systems, which often transcend straightforward causal relationships.

To operationalise this approach, causal loop diagrams (CLDs) were constructed using Vensim Software. These diagrams were instrumental in visually mapping out the relationships and feedback mechanisms identified in the literature. Through thematic analysis of the literature, key variables influencing ethical and privacy considerations were identified. These variables served as the basis for the CLDs, enabling a detailed exploration of how changes in one aspect of the system might ripple through others, influencing the overall functionality and ethical alignment of AI-enabled syndromic surveillance systems.

The use of systems dynamic modeling (SDM), specifically through the development of CLDs, provided a structured method to analyse the complexity of these systems. It facilitated a deeper understanding of potential leverage points and areas of vulnerability, particularly in the context of under-resourced settings in Southern Africa. This comprehensive approach ensured a holistic analysis, capturing the delicate and often indirect effects of various factors on the ethical and privacy dimensions of AI-enabled syndromic surveillance. The study's focus on South Africa and Zimbabwe offers specific insights while contributing to the broader discourse on responsible AI in healthcare.

Causal Loop Diagram Development. The development of Causal Loop Diagrams (CLDs) assists us in comprehending and illustrating the feedback processes created in intricate systems. These diagrams capture the connections, behaviors, and time lags among variables that form these systems [20,21,39]. In essence, they provide a pragmatic avenue to grasping and articulating the interconnected components of the system and the causal relationships relevant to the specific issue being addressed and, in this case, ethical and privacy considerations in AI-enabled syndromic surveillance systems [18,40]. CLDs constitute two key elements: variables and influences, or links. Each influence is represented by an arrow to indicate its direction, along with a sign that shows whether the element being influenced changes in the same direction (+) or the opposite direction (−) as the element exerting the influence.

Feedback loops arise in situations where a variable is connected back to itself via a sequence of other variables within Causal Loop Diagrams (CLDs). There are two distinct categories of feedback loops utilised in CLDs: balancing and reinforcing loops.

Balancing loops are prevalent in scenarios aiming for problem resolution or goal attainment. Occasionally, these are termed neutralizing loops, characterized by cause-and-effect sequences that strive to counterbalance a change with an opposing force. An example (see Fig. 1) illustrates the relationship amongst variables, "In-formed Consent", "Community Engagement and Trust", and "Algorithmic Fairness and Bias". In the balancing loop context, enhanced informed consent boosts community trust, which can help in addressing algorithmic biases. In turn, the awareness of these biases can lead to more stringent consent requirements, creating a self-regulating system.

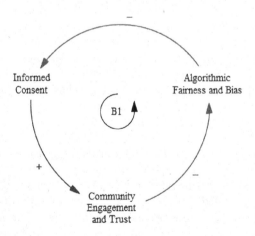

Fig. 1. Balancing loop example.

On the other hand, reinforcing loops depict escalating actions, where each successive action amplifies the previous one. These loops can be classified as virtuous cycles when they yield positive outcomes or as vicious cycles when they result in detrimental consequences. Figure 2 illustrates the relationship amongst variables, "Transparency and Accountability", "Community Engagement and Trust", and "Beneficence and Nonmaleficence" in drawing up a reinforcing loop. In this reinforcing loop, transparency and accountability foster ethical practices, which build community trust. In turn, increased trust drives a demand for even more transparency and accountability, creating a cycle that continuously strengthens these ethical principles.

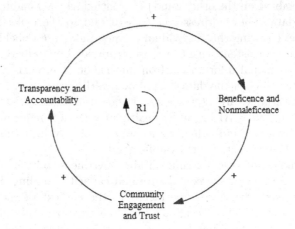

Fig. 2. Reinforcing loop example.

In summary, these loops demonstrate how different variables in the ethical framework for AI-enabled syndromic surveillance systems are interconnected, influencing each other in complex ways. The balancing loop shows a self-regulatory mechanism, while the reinforcing loop indicates a cycle of continuous improvement.

Causal Loop Diagram Validation. Review and validation of the CLDs were conducted by the authors of this study together with other externally identified researchers in AI and healthcare. This exercise was deemed necessary to ensure we had included most of the highly relevant factors and correct relationships created. In the case where we had missing factors, we would include them and revise the diagram. Furthermore, reviewers examined the correctness of the directional orientation of each link, assessing if the supposed effects were the causes, thus requiring a reversal of the links. They were also prompted to identify any additional outcomes that might emerge as a consequence of the causes depicted in the Causal Loop Diagrams (CLDs). Suggested inputs from reviewers were then discussed by the authors of this paper and then used to improve the CLDs while also consulting evidence from the literature.

3   Findings

Our findings were split according to two sets of CLD variable dynamics, that is, ethical and privacy variables affecting AI. In the context of AI-enabled syndromic surveillance, ethical variables primarily revolve around the moral principles guiding system development and deployment, focusing on fairness, transparency, community engagement, and regulatory alignment. These considerations are crucial to ensure that such systems are used responsibly, respecting human dignity and societal norms.

Privacy variables, on the other hand, are centred on the handling of personal and sensitive data. They address concerns related to the collection, security, and use of data, ensuring that individual privacy rights are upheld, and data is used in a manner consistent with the consent provided by individuals. The goal is to safeguard personal information from misuse or unauthorised access while maintaining the utility of the data for public health purposes.

Both sets of variables are interdependent and vital for the ethical and privacy-conscious deployment of AI in public health surveillance, requiring a balanced approach that respects individual rights and societal values while leveraging technological advancements for the public good.

There are, however, cross-cutting variables pertinent to both ethics and privacy and these are regulatory compliance and capacity building. Each of these variables and their drivers was considered in the context of their impact on both privacy and ethics when designing, implementing, or regulating AI-enabled syndromic surveillance systems. Key findings are presented as follows:

3.1   Ethical and Privacy Considerations in AI-Enabled Disease Surveillance

Algorithmic Bias and Fairness. Algorithmic bias and fairness are crucial in AI ethics for disease surveillance due to their potential to perpetuate inequalities and reinforce harmful stereotypes [41]. In healthcare, these biases can lead to disparities in diagnosis, treatment, and healthcare costs, particularly across race sub-populations [10]. The use of AI in disease surveillance, therefore, requires careful consideration of these issues to ensure equitable care. Mitigation strategies such as diverse and representative datasets, enhanced transparency and accountability, and alternative AI paradigms prioritizing fairness and ethical considerations are essential [41]. Additionally, the use of emerging technologies like federated learning, disentanglement, and model explainability can help mitigate bias in AI systems [10,42,43]. Diversity in algorithm development teams can help in recognising and mitigating biases that might otherwise be overlooked.

Community Trust and Engagement. Community trust is essential for the successful implementation and effectiveness of AI-enabled surveillance systems. Engagement with community stakeholders fosters trust and helps in tailoring the system to local contexts and needs [11,44]. Trust and social relationships are essential for ethical good practice in global health. Efforts should be made to strengthen responsible AI frameworks that underscore the importance of transparent communication and inclusive practices within communities [45,46].

Transparency and Accountability. Transparency in AI-enabled surveillance systems and accountability for their outcomes are necessary to build and maintain public trust. This involves clear communication about how the systems work, their limitations, and who is responsible for decisions. Studies suggest public demand and regulatory requirements as driving forces for greater transparency [26,38,47,48]. These concerns are particularly relevant in the context of infectious diseases, where the use of AI can have significant implications for public health and human rights [49]. Therefore, ensuring transparency and accountability in the development and deployment of AI technologies for disease surveillance is essential to build trust and mitigate potential harm.

Surveillance Scope and Proportionality. Balancing surveillance scope with the need to respect individual privacy and freedom is key. Surveillance should be proportionate to public health goals, ensuring it's not overly intrusive [44,47]. This therefore underscores the need for a balanced approach to surveillance in AI ethics, one that ensures the protection of individual rights and privacy while also enabling effective disease monitoring and response [11].

Autonomy and Individual Rights. Respecting individual rights and autonomy is fundamental in ethical surveillance. This includes safeguarding personal freedoms and respecting individuals' decisions regarding their data [11,44,50]. Societal values and norms play a significant role in determining the extent of these rights. The potential for AI to challenge human autonomous decisions in disease surveillance further underscores the importance of these considerations [22,43,51].

Long-Term Impacts and Unintended Consequences. The long-term impacts and unintended consequences of AI in disease surveillance are crucial considerations in AI ethics as they pose a potential for AI-enabled surveillance to exacerbate existing health disparities and infringe on privacy and civil rights [11,38,44]. Policymakers and regulators need to address ethical risks in AI health care, including those related to fairness and transparency [46]. The potential for these issues to erode public trust in AI for health care is a significant long-term consequence that must be carefully managed.

Data Collection and Consent. Data collection and consent are a cornerstone in AI ethics for disease surveillance due to the potential for privacy violations and the exacerbation of existing health disparities. The use of AI in disease surveillance, while promising, also raises complex ethical issues, including the potential for human autonomous decisions to be challenged [11, 38, 43, 44, 48, 49, 52]. Therefore, ensuring that data collection is conducted ethically and with the consent of individuals is essential to mitigate these risks and uphold the rights of individuals, particularly in vulnerable and at-risk populations.

Data Security and Protection. Data security and protection are crucial in AI ethics for disease surveillance due to the potential for misuse and infringement of individual rights. Studies have raised ethical concerns about privacy, confidentiality, and autonomy in the use of AI-augmented surveillance systems for infectious disease detection [36, 44, 47, 49, 50]. There is a need for emphasis on responsible data use and advances in cybersecurity in AI-enabled disease surveillance [11, 49, 52]. The potential for data exposure and misuse in disease surveillance underscores the importance of robust data security and protection measures to ensure the ethical use of AI in this context.

Data Utilisation and Purpose Limitation. Surveillance data utilization and purpose limitation are crucial in AI ethics for disease surveillance due to their potential impact on individual dignity, civil liberties, and transparency [49]. The responsible use of AI in disease surveillance, particularly in the context of global health, is a key consideration, with a need for a framework that addresses the complexity and potential for human autonomous decisions [11]. The ethical concerns raised by digital technologies and new data sources in public health surveillance, such as uneven distribution of trust and participation, and privacy concerns, further underscore the importance of these principles [48].

Data Sharing and Accessibility. There are several reasons surveillance data sharing and accessibility are essential for the ethical and effective use of AI-enabled disease surveillance systems. Firstly, it enables the integration of various early warning tools, enhancing risk assessment and response [52]. Secondly, it supports the responsible use of AI in disease surveillance, ensuring digital integrity, trust, and privacy protection [44]. Thirdly, it allows for the collection and analysis of vast amounts of real-time data, informing epidemiological and public health emergency responses [11]. Lastly, it addresses the ethical concerns raised by the use of digital technologies in public health surveillance, such as uneven distribution of trust and participation, and privacy issues [48].

Anonymisation and De-identification. Anonymisation techniques are vital for using data in surveillance without compromising the balance between data utility with individual privacy [53]. The use of AI in disease surveillance, particularly in the context of global health, has the potential to collect and analyze vast amounts of data but also raises complex ethical issues [11]. Anonymisation through data synthesis can provide a solution to these issues by generating synthetic data that closely approximates the original data, thus enabling more open data sharing while minimising the risk of breaching patient confidentiality [54]. Technological advancements in this field enhance the ability to use data safely.

There are other cross-cutting variables, that is, variables that are crucial and exist in both the ethics and privacy contexts of AI considerations. These include regulatory frameworks and capacity building.

Regulatory Frameworks. Robust regulatory and legislative frameworks form the backbone of ethics and privacy considerations in AI for disease surveillance due to the potential for ethical issues and human rights violations. There is a need for clear governance frameworks and responsible AI frameworks to address concerns such as privacy protection, data integrity, and autonomy [11,22,44,48]. These frameworks can help ensure that vulnerable populations are not under-represented and that highly sensitive data is protected [9,10,35,36,38,51]. It is therefore important to emphasise a responsible AI framework in addressing the complex and dynamic nature of AI-enabled disease surveillance.

Capacity Building and Education. Educating stakeholders about ethical and privacy considerations in AI is crucial for the responsible deployment and operation of surveillance systems due to the complex nature of ethical issues that arise from them [10,11,24,44]. Studies collectively underscored the necessity of investment in training and development as essential for building this capacity and creating a knowledge base amongst all concerned stakeholders to navigate these ethical challenges while ensuring that AI-enabled disease surveillance is conducted responsibly and ethically [46,47,50].

4   Causal Loop Diagrams

Two CLDs depicting the variables associated with ethics (see Fig. 3), and privacy (see Fig. 4) considerations for AI-enabled syndromic surveillance systems were created from the literature reviewed together with brainstorming sessions among the study authors. Several reinforcing and balancing feedback loops can be observed in these CLDs. A detailed analysis of the CLDs is provided below:

ETHICAL CONSIDERATIONS

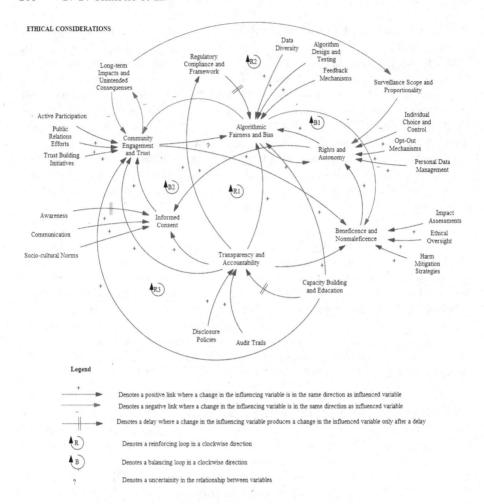

Fig. 3. Ethical considerations causal loop diagram.

CLD Dynamics in AI Syndromic Surveillance Ethics. The dynamics involved in variables under ethical considerations are presented in Fig. 3. We identified two balancing loops (B1) where there is an attempt to achieve fairness and equity, and (B2) where there is an attempt to emphasise the importance of trust building in AI ethics. There were also three reinforcing loops (R1) - ethical in AI development loop; (R2) - a vicious cycle of bias and distrust loop; and (R3) - community empowerment loop as explained below:

In the balancing loop (B1), the interaction among "Beneficence and Non-maleficence," "Rights and Autonomy," and "Algorithmic Fairness and Bias" reflects a critical ethical equilibrium in AI-enabled syndromic surveillance systems. The pursuit of beneficence (doing good) and non-maleficence (avoiding harm) positively contributes to respecting individual rights and autonomy,

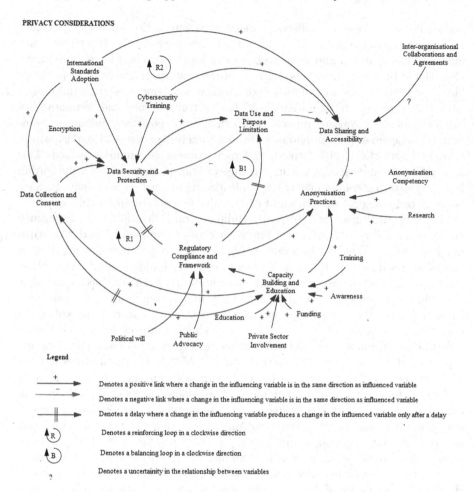

PRIVACY CONSIDERATIONS

Legend

+ →	Denotes a positive link where a change in the influencing variable is in the same direction as influenced variable
− →	Denotes a negative link where a change in the influencing variable is in the same direction as influenced variable
‖→	Denotes a delay where a change in the influencing variable produces a change in the influenced variable only after a delay
R	Denotes a reinforcing loop in a clockwise direction
B	Denotes a balancing loop in a clockwise direction
?	Denotes a uncertainity in the relationship between variables

Fig. 4. Privacy considerations causal loop diagram.

ensuring that the system's operations do not infringe on personal freedoms. However, the presence of algorithmic bias negatively impacts the principles of beneficence and nonmaleficence, as biased outcomes can inadvertently cause harm or unfair treatment. Simultaneously, the respect and maintenance of rights and autonomy positively influence efforts to reduce algorithmic bias. This positive influence suggests that a strong commitment to individual rights fosters an environment where biases are more diligently identified and addressed. This loop represents the delicate balance between the ethical use of AI (beneficence and nonmaleficence), the preservation of individual rights and autonomy, and the continuous effort to mitigate algorithmic bias, highlighting the interconnectedness and potential tensions between these critical ethical considerations.

The balancing loop (B2) with variables "Transparency and Accountability," "Algorithmic Fairness and Bias," "Community Engagement and Trust," and "Informed Consent" reflects a meticulous interaction of ethical dynamics in emphasising trust in AI-enabled syndromic surveillance. It suggests that increased transparency and accountability can lead to a greater focus on reducing algorithmic bias, enhancing the fairness of the system. However, if algorithmic bias is present, it negatively impacts community engagement, as trust diminishes when bias is perceived. Simultaneously, greater transparency and accountability enhance informed consent processes, which in turn positively influences community engagement, as individuals are more likely to trust and participate in systems where they feel informed, and their consent is genuinely valued. This loop therefore underscores the importance of transparency in both mitigating bias and fostering community trust, while also highlighting the delicate balance between technological integrity and community perception and involvement.

The loop (R1) involving "Capacity Building and Education," "Transparency and Accountability," "Regulatory Framework and Compliance," and "Algorithmic Fairness and Bias" forms a reinforcing feedback loop, although with delayed effects in certain connections. In this loop, capacity-building efforts, over time, lead to enhanced transparency and accountability in AI-enabled syndromic surveillance systems. Increased transparency and accountability, in turn, positively influence the development and enforcement of regulatory frameworks and compliance mechanisms. These frameworks, though subject to delayed implementation, ultimately contribute to reducing algorithmic bias. Furthermore, capacity building directly impacts the reduction of algorithmic bias by educating and equipping stakeholders with the necessary skills and knowledge to identify and mitigate biases in AI development. The delayed effects in the loop indicate that improvements in transparency, accountability, and regulatory compliance don't happen instantaneously but unfold over time. The overall reinforcing nature of the loop suggests a cumulative, positive impact on ethical AI practices, where initial investments in capacity building set in motion a series of improvements that cyclically enhance the system's ethical standards and effectiveness. However, the presence of delays accentuates the need for patience and sustained effort, as the benefits of such initiatives may only become evident in the longer term.

In this reinforcing loop (R2), the interplay of variables creates a cycle that can exacerbate certain challenges in AI-enabled syndromic surveillance. Algorithmic bias negatively impacts community engagement, as communities tend to disengage when they perceive bias. This reduced engagement can lead to more significant long-term impacts and unintended consequences, as community feedback is crucial for identifying and mitigating such issues. These consequences can then necessitate a reduction in surveillance scope, often as a response to protect public trust or due to regulatory actions. A reduced surveillance scope, in turn, negatively impacts the system's ability to respect rights and autonomy, potentially leading to more stringent privacy protections or limitations. Interestingly, this heightened focus on rights and autonomy may lead to increased efforts to address algorithmic bias, as systems aim to regain community trust. However, this can inadvertently perpetuate the cycle, as efforts to reduce bias might not fully align with community expectations or fail to address underlying issues, leading again to decreased community engagement. In summary, this loop highlights the complex and sometimes counter-intuitive interactions between community perceptions, system adjustments, and ethical considerations, illustrating how well-intentioned changes can perpetuate underlying challenges.

This reinforcing loop(R3) centred around "Capacity Building and Education," "Transparency and Accountability," and "Community Engagement and Trust," demonstrates the positive cycle initiated by capacity building in AI-enabled syndromic surveillance systems. Capacity building, though with a delayed effect, enhances transparency and accountability, as better-trained personnel are more adept at ensuring these values in system operations. This increase in transparency and accountability, in turn, positively influences community engagement, as communities tend to trust and engage more with systems that they perceive as transparent and accountable. Simultaneously, capacity building directly contributes to enhanced community engagement, as stakeholders, including community members, become more informed and capable of engaging with the system. This loop emphasises the role of education and skill development in fostering a more engaged, informed, and trusting community, essential for the successful implementation and acceptance of surveillance systems.

There are variables in the CLD such as the relationship between "Community Engagement and Trust" and "Algorithmic Fairness and Bias," indicated with a question mark (?), which suggests an area of uncertainty or complexity in understanding how these variables interact in the context of AI-enabled syndromic surveillance.

One possible lens to this is that increased community engagement and trust could lead to improvements in algorithmic fairness and bias. When communities are more engaged and trust the system, they might provide more diverse data and feedback, which can be used to identify and correct biases in the algorithms. Additionally, high levels of trust and engagement might pressure developers and policymakers to pay closer attention to fairness issues, leading to more equitable algorithms. Conversely, it is also conceivable that despite high community engagement and trust, algorithmic biases might not be effectively addressed due to other overriding factors like technical limitations, lack of expertise, or institutional inertia.

In summary, the "?" symbol therefore represents this uncertainty or the multiple potential directions of influence. It indicates a need for further investigation to understand how community engagement and trust might influence algorithmic fairness and bias, acknowledging that the relationship is not straightforward and could be influenced by various external factors.

CLD Dynamics in AI Syndromic Surveillance Privacy. The dynamics involved in variables under privacy considerations are presented in Fig. 4. We identified one balancing loop (B1) - data privacy management loop and two reinforcing loops (R1) - privacy empowerment cycle loop; (R2) - data security enhancement loop as explained below:

In this balancing loop (B1) within the privacy considerations of AI-enabled syndromic surveillance, there's an intricate interaction between "Data Sharing and Accessibility," "Anonymisation Practices," "Data Security and Protection," and "Data Use and Purpose Limitation." Increased data sharing and accessibility can compromise anonymisation practices, as broader data distribution makes it challenging to maintain strict anonymisation. However, effective anonymisation practices positively contribute to data security and protection, as they safeguard personal information even when data breaches occur. On the other hand, stringent data use and purpose limitations, while essential for privacy, can paradoxically reduce data security. This happens because overly restrictive data use limits the sharing of potentially useful information for enhancing security measures. Furthermore, strict purpose limitations also hamper data sharing and accessibility, creating a tension between ensuring privacy and facilitating the beneficial use of data. This loop highlights the delicate balance needed in managing data sharing, anonymisation, security, and usage limitations to maintain both privacy and functionality in surveillance systems.

This reinforcing loop (R1) within the privacy considerations of AI-enabled syndromic surveillance systems highlights a progressive cycle that starts with "Data Collection and Consent." Improved practices in data collection and obtaining consent, though with a delayed effect, contribute to enhancing "Capacity Building and Education." As stakeholders become more educated and skilled, this leads to better "Regulatory Compliance and Framework," ensuring that data management adheres to established privacy and security standards. However, this compliance also has a delayed impact on "Data Security and Protection," as the implementation of regulations and their integration into practices take time. Simultaneously, improvements in data collection and consent directly feed into enhancing data security and protection. This loop demonstrates a virtuous cycle where informed data collection practices and consent processes lead to better education and understanding, which in turn improves regulatory compliance. This compliance, albeit delayed, ultimately reinforces data security, creating a continually improving system of data protection.

This reinforcing loop (R2), involving "Cybersecurity Training," "Data Security and Protection," and "International Standards Adoption," illustrates a positive feedback cycle enhancing data security and accessibility in AI-enabled syndromic surveillance. Enhanced cybersecurity training boosts data security, equipping personnel with the skills to protect sensitive information. Similarly, adopting international standards not only bolsters data security but also facilitates data sharing by providing a common framework and protocols. This shared framework, in turn, encourages more data sharing, as entities are more confident in the security and compatibility of shared data. Moreover, the knowledge and practices gained from cybersecurity training further support secure and efficient data sharing. Overall, this loop highlights how training and standardisation work hand in hand to improve both data security and the ease of data exchange, creating a virtuous cycle of continuous improvement in data management.

There were variable relationships in the CLD which we thought had some uncertainties. As such, the relationship between "Inter-organisational Collaborations and Agreements" and "Data Sharing and Accessibility," marked with a question mark (?), indicates a complex and potentially uncertain connection in the context of AI-enabled syndromic surveillance systems. On one hand, collaborations and agreements between organisations could facilitate greater data sharing and accessibility, as they might establish common standards and protocols, thereby making it easier to share data securely and efficiently. However, the actual impact of these collaborations on data sharing could vary. Factors such as the nature of the agreements, the level of trust between organisations, and the technological compatibility of different systems could significantly influence how effectively data is shared. The relationship suggests that while inter-organisational collaborations have the potential to enhance data sharing, the extent and effectiveness of this enhancement are not always straightforward and can depend on several subtle factors.

5 Discussion

In this study, we present an intensive complex analysis of the interactions of different variables towards ethical and privacy considerations when implementing AI-enabled syndromic surveillance systems in Southern Africa. An examination of feedback loops constructed due to the complex and dynamic stature of ethics in AI was conducted. This assisted us in the determination and generation of strategy pointers that may be used towards addressing ethical and privacy challenges when implementing AI-enabled syndromic surveillance systems in resource-constrained contexts such as Southern Africa.

A crucial strength of this study is the deduction that AI-enabled surveillance systems are tools developed and deployed to uphold the wellness of society. The emphasis placed on the inclusion of perspectives from different stakeholders (policy-makers, researchers, developers, health professionals, community leaders, and ordinary citizens) in addressing ethical and privacy concerns related to AI has played a crucial role in demystifying the complex and dynamic interactions of different variables in AI and its application in public health surveillance for resilient health systems [21, 40]. We also managed to engage other researchers in AI and health to validate our CLDs, which helped address some of the shortcomings in the models to strengthen our argument.

This study had its limitations as it focused on the two countries (South Africa and Zimbabwe) in the Southern African region with different socio-economic and political contexts. Generalisation of the findings to the Southern African Region can be a problem due to these differences and, our models were developed using insights drawn from mostly developed countries that have developed and deployed AI-enabled syndromic surveillance systems. However, it is crucial to note that most of the critical variables are universal and can be applied and altered to suit any context.

5.1 High-Level Leverage Points

There are several leverage points identified from our findings that yield significant positive impacts and enhance both the ethical integrity and privacy protection of the surveillance system. Some of these high-leverage points include:

Transparency and Public Engagement: There is a need to actively engage with and inform the public about how surveillance data is used, addressing privacy, data integrity, and governance, and being transparent about the purposes and limitations of the surveillance can build trust and compliance [44,48,51]. Transparency is also envisaged in the design of the systems and the level of participation from the community. Involving community representatives in the design and implementation phases of the surveillance system ensures that the system aligns with the values, needs, and expectations of the community, thereby enhancing its ethical acceptability is important. Challenges such as potential bias and discriminatory outcomes are averted using such a community-centric design approach [11,43,44,49,55].

Algorithmic Oversight and Bias Mitigation: A key point in our findings affirms the need to implement regular audits and reviews of AI algorithms for potential biases and develop a protocol for bias mitigation [42]. Tying into the loop between algorithmic bias, community trust, and systems effectiveness, focusing on reducing bias in AI algorithms can improve both ethical outcomes and systems effectiveness. There is an emphasis on healthcare organisations adopting policies and practices that ensure accuracy, fairness, and lack of bias in AI algorithms [56,57].

Regulatory Compliance and Framework: Leveraging the loop between regulatory compliance, data security, and public trust, studies suggest that strengthening the legal and regulatory framework that governs AI-enabled surveillance systems, ensuring it is up-to-date with technological advancements, can be a high-impact area [44]. The use of AI in disease surveillance presents significant ethical challenges, particularly in the areas of privacy, data integrity, and governance [22,35,52]. These challenges are further compounded by the need for data sharing and the regulation of public health data. The COVID-19 pandemic presented itself as a good example as it accelerated the use of AI-assisted surveillance technologies, raising concerns about individual dignity, civil liberties, and data protection [49,58]. There is a need to establish robust data protection policies that encompass consent, collection, use, and sharing of data can significantly enhance privacy. This includes strict enforcement of data anonymisation and de-identification practices. Currently, the Health Acts and Protection of Information Acts of the two cited countries haven't yet been updated to address the issue of AI ethics application in any domain [59–63].

Capacity Building and Education Investment. Building capacity among stakeholders, especially in understanding AI, data privacy, and ethical implications, is pivotal. By educating those involved in the development and implementation of AI systems, biases can be reduced, and ethical considerations can be more effectively integrated [22,35,37,41,44,64]. This leverages the loop connecting capacity building, ethical practices, and community trust.

Feedback Mechanisms and Collaborations: Establishing robust feedback loops within the system to continuously monitor, evaluate, and respond to ethical and privacy concerns can make the system more adaptable and responsive to emerging challenges [11,42,48]. Strengthening collaborations with diverse stakeholders, including international bodies, for sharing best practices, resources, and knowledge can help in maintaining high standards of privacy and ethical considerations, which is crucial [11,49].

6   Conclusions

This study using the systems thinking approach, systematically analysed the ethical and privacy considerations in AI-enabled syndromic surveillance systems, particularly in resource-constrained regions such as Southern Africa. We identified crucial variables such as algorithmic bias and fairness, community trust and engagement, data security, and informed consent, each playing a significant role in shaping the effectiveness and ethical integrity of these systems. Our findings highlight a complex interplay between ethical imperatives and privacy concerns, emphasising the need for a balanced approach that respects individual rights while leveraging the benefits of AI for public health.

The causal loop diagrams developed in this study illustrate reinforcing and balancing feedback loops, underscoring the dynamic nature of these systems. Key leverage points, including enhancing transparency and accountability, strengthening regulatory frameworks, and focusing on capacity building, emerged as critical for fostering trust and ensuring privacy and ethical compliance.

In Southern Africa, a region with limited resources, the challenges are compounded by infrastructural constraints, varied cultural perceptions of privacy, and differing levels of technological adoption. However, these challenges also present opportunities for innovative, context-specific solutions that respect local norms and values while harnessing the power of AI for public health surveillance. Areas of future research based on this study may include studies on, ethical AI design and development models, the impact of regulatory frameworks in AI, and community-centric design approaches amongst others.

References

1. United Nations: Transforming Our World: The 2030 Agenda for Sustainable Development, p. 41. UN, Geneva (2015)
2. WHO: Global strategy on digital health 2020–2025, WHO, Editor, p. 41. WHO, Switzerland (2019)

3. WHO: 13th General Programme of Work 2019–2023, WHO, Editor, p. 60. WHO, Switzerland (2018)
4. Dey, K.: Digital Health Market Research Report, p. 210. Market Research Future, UK (2021)
5. Statista: Digital Health - South Africa, Statista (2021)
6. Statista: Digital Health - Zimbabwe. Statista (2021)
7. insights10.: Market Research Report: South Africa Digital Health market Analysis, V. Upadhyay, Editor, India (2023)
8. Craven, M., Sabow, A., Van der Veken, L., Wilson, M.: Not the last pandemic: investing now to reimagine public-health systems, M. Company, Editor. McKinsey & Company (2020)
9. Baclic, O., et al.: Artificial intelligence in public health: challenges and opportunities for public health made possible by advances in natural language processing. Can. Commun. Dis. Rep. **46**(6), 161 (2020)
10. Chen, H., et al.: AI for global disease surveillance. IEEE Intell. Syst. **24**(6), 66–82 (2009)
11. Borda, A., et al.: Ethical issues in AI-enabled disease surveillance: perspectives from global health. Appl. Sci. **12**(8), 3890 (2022)
12. Coovadia, H., et al.: The health and health system of South Africa: historical roots of current public health challenges. The Lancet **374**(9692), 817–834 (2009)
13. Gilson, L., Daire, J.: Leadership and governance within the South African health system. S. Afr. Health Rev. **2011**(1), 69–80 (2011)
14. Loewenson, R., Sanders, D., Davies, R.: Challenges to equity in health and health care: a Zimbabwean case study. Soc. Sci. Med. **32**(10), 1079–1088 (1991)
15. Schneider, H., et al.: Health systems and access to antiretroviral drugs for HIV in Southern Africa: service delivery and human resources challenges. Reprod. Health Matters **14**(27), 12–23 (2006)
16. Witter, S., et al.: The political economy of results-based financing: the experience of the health system in Zimbabwe. Glob. Health Res. Policy **4**, 1–17 (2019)
17. Fontes, C., et al.: AI-powered public surveillance systems: why we (might) need them and how we want them. Technol. Soc. **71**, 102137 (2022)
18. Adam, T., de Savigny, D.: Systems thinking for strengthening health systems in LMICs: need for a paradigm shift. Health Policy Plan. **27**(suppl4), iv1–iv3 (2012)
19. Fong, B.Y.: Systems Thinking and Sustainable Healthcare Delivery. Taylor & Francis, Routledge (2022)
20. Leischow, S.J., Milstein, B.: Systems thinking and modeling for public health practice, pp. 403–405. American Public Health Association (2006)
21. Strachna, O., Asan, O.: Systems thinking approach to an artificial intelligence reality within healthcare: from hype to value. In: 2021 IEEE International Symposium on Systems Engineering (ISSE). IEEE (2021)
22. World Health Organisation: Regulatory considerations on artificial intelligence for health, p. 61. World Health Organisation, Geneva (2023)
23. Allam, Z., Dey, G., Jones, D.S.: Artificial intelligence (AI) provided early detection of the coronavirus (COVID-19) in China and will influence future Urban health policy internationally. Ai **1**(2), 156–165 (2020)
24. Brownstein, J.S., et al.: Advances in Artificial Intelligence for infectious-disease surveillance. N. Engl. J. Med. **388**(17), 1597–1607 (2023)
25. Pathman, A., et al.: Knowledge, attitudes, practices and health beliefs toward leptospirosis among urban and rural communities in northeastern Malaysia. Int. J. Environ. Res. Public Health **15**(11), 2425 (2018)

26. Boch, A., Corrigan, C.: Ethics and the use of AI-based tracing tools to manage the COV ID-19 pandemic. TUM IEAI Research Brief. (2020)
27. Rohmetra, H., et al.: AI-enabled remote monitoring of vital signs for COVID-19: methods, prospects and challenges. Computing, 1–27 (2021)
28. Ward, R.J., et al.: FluNet: an AI-enabled influenza-like warning system. IEEE Sens. J. **21**(21), 24740–24748 (2021)
29. Bozzola, E., et al.: Global measles epidemic risk: current perspectives on the growing need for implementing digital communication strategies. Risk Manag. Healthc. Policy **13**, 2819–2826 (2020)
30. Nazir, A., et al.: Upsurge of measles in South Africa: A cause for concern? (2023)
31. Chimusoro, A., et al.: Responding to cholera outbreaks in Zimbabwe: building resilience over time. Current Issues Global Health, pp. 45–64 (2018)
32. Mavhunga, C.: Cholera: World Health Organization warns of rising cases in Africa. Br. Med. J. Publ. Group **380**, p488 (2023)
33. Smith, A.M., et al.: Imported Cholera Cases, South Africa, 2023. Emerg. Infect. Dis. **29**(8), 1687 (2023)
34. Sikhosana, M.L., Kuonza, L., Motaze, N.V.: Epidemiology of laboratory-confirmed mumps infections in South Africa, 2012–2017: a cross-sectional study. BMC Publ. Health **20**(1), 1–9 (2020)
35. Corrêa, N.K., et al.: Worldwide AI ethics: a review of 200 guidelines and recommendations for AI governance. Patterns 4(10), 100857 (2023)
36. Lee, R.S., Lee, R.S.: AI ethics, security and privacy. Artif. Intell. Daily Life, 369–384 (2020)
37. Jobin, A., Ienca, M., Vayena, E.: The global landscape of AI ethics guidelines. Nat. Mach. Intell. **1**(9), 389–399 (2019)
38. Hermansyah, M., et al.: Artificial intelligence and ethics: building an artificial intelligence system that ensures privacy and social justice. Int. J. Sci. Soc. **5**(1), 154–168 (2023)
39. Jalali, M.S., Beaulieu, E.: Strengthening a weak link: transparency of causal loop diagrams-current state and recommendations. Syst. Dyn. Rev. (2023)
40. De Savigny, D., Blanchet, K., Adam, T.: EBOOK: Applied Systems Thinking for Health Systems Research: A Methodological Handbook. McGraw-Hill Education (UK), Udgiver (2017)
41. Ferrara, E.: Fairness and bias in artificial intelligence: a brief survey of sources, impacts, and mitigation strategies. arXiv preprint arXiv:2304.07683 (2023)
42. Fletcher, R.R., Nakeshimana, A., Olubeko, O.: Addressing fairness, bias, and appropriate use of artificial intelligence and machine learning in global health. Frontiers Media SA, p. 561802 (2023)
43. Zhang, J., Shu, Y., Yu, H.: Fairness in design: a framework for facilitating ethical artificial intelligence designs. Int. J. Crowd Sci. **7**(1), 32–39 (2023)
44. Zhao, I.Y., et al.: Ethics, integrity, and retributions of digital detection surveillance systems for infectious diseases: systematic literature review. J. Med. Internet Res. **23**(10), e32328 (2021)
45. Adhikari, B., Pell, C., Cheah, P.Y.: Community engagement and ethical global health research. Global Bioethics **31**(1), 1–12 (2020)
46. Morley, J., et al.: The ethics of AI in health care: a mapping review. Soc. Sci. Med. **260**, 113172 (2020)
47. Li, F., Ruijs, N., Lu, Y.: Ethics & AI: a systematic review on ethical concerns and related strategies for designing with AI in healthcare. AI **4**(1), 28–53 (2022)
48. Mello, M.M., Wang, C.J.: Ethics and governance for digital disease surveillance. Science **368**(6494), 951–954 (2020)

49. Findlay, M., et al.: Ethics, AI, mass data and pandemic challenges: responsible data use and infrastructure application for surveillance and pre-emptive tracing post-crisis. SMU Centre for AI & Data Governance Research Paper (2020)
50. World Health Organization: Ethics and governance of artificial intelligence for health: WHO guidance (2023)
51. Al-Hwsali, A., et al.: Scoping review: legal and ethical principles of artificial intelligence in public health. Stud. Health Technol. Inform. **305**, 640–643 (2023)
52. Kostkova, P.: Disease surveillance data sharing for public health: the next ethical frontiers. Life Sci. Soc. Policy **14**(1), 1–5 (2018)
53. Gkoulalas-Divanis, A., Loukides, G., Sun, J.: Toward smarter healthcare: anonymizing medical data to support research studies. IBM J. Res. Dev. **58**(1), 9: 1–9: 11 (2014)
54. Yoon, J., Drumright, L.N., Van Der Schaar, M.: Anonymization through data synthesis using generative adversarial networks (ads-GAN). IEEE J. Biomed. Health Inform. **24**(8), 2378–2388 (2020)
55. Gerdes, A.: A participatory data-centric approach to AI Ethics by Design. Appl. Artif. Intell. **36**(1) (2022)
56. Delgado, J., et al.: Bias in algorithms of AI systems developed for COVID-19: a scoping review. J. Bioethical Inq. **19**(3), 407–419 (2022)
57. McCall, C.J., DeCaprio, D., Gartner, J.: The Measurement and Mitigation of Algorithmic Bias and Unfairness in Healthcare AI Models Developed for the CMS AI Health Outcomes Challenge. medRxiv (2022)
58. Su, Z., Bentley, B., Shi, F.: Artificial intelligence-based disease surveillance amid COVID-19 and beyond: a systematic review protocol (2020)
59. Postal and Telecommunications Regulatory Authority: Data Protection Act, POTRAZ, Editor, p. 38. Government of Zimbabwe, Zimbabwe (2021)
60. Ministry of Information and Publicity: Access to Information and Protection of Privacy Act, I.a. Publicity, Editor, p. 54. Government of Zimbabwe, Harare, Zimbabwe (2023)
61. Ministry of Health Zimbabwe: Public Health Act, H. Department, Editor, p. 76. Veritas, Harare, Zimbabwe (2018)
62. Government of South Africa: National Health Act, Health, Editor, p. 48. Government of South Africa, South Africa (2004)
63. Government of South Africa: Protection of Personal Information Act, Information, Editor, p. 76. Government of South Africa, South Africa (2013)
64. Munn, L.: The uselessness of AI ethics. AI Ethics **3**(3), 869–877 (2023)
65. Van Zyl, C., Badenhorst, M., Hanekom, S., Heine, M.: Unravelling 'low-resource settings': a systematic scoping review with qualitative content analysis. BMJ Global Health **6**(6), e005190 (2021)
66. Gibbons, C.L., et al.: Measuring underreporting and under-ascertain. BMC Publ. Health **14**(1), 1–17 (2014)

Fake vs. Real Face Discrimination Using Convolutional Neural Networks

Khaled Eissa[1] and Friedhelm Schwenker[2]([✉])

[1] German University in Cairo, New Cairo, Egypt
khaled.anwar@student.guc.edu.eg
[2] Institute of Neural Information Processing, Ulm University, Ulm, Germany
friedhelm.schwenker@uni-ulm.de

Abstract. The rapid advancements in image manipulation technology and the proliferation of Generative Adversarial Network (*GAN*) generated content have created a pressing need for effective methods to distinguish between fake and real imagery. Convolutional Neural Networks (*CNNs*) have exhibited exceptional performance in image recognition tasks, making them an ideal choice for addressing the challenge of fake vs. real face discrimination. This paper proposes three new *CNN* models specifically designed for discriminating between fake and real facial images. The proposed models are compared with several state-of-the-art pre-trained models, such as Inception-ResNet-V2. An extensive dataset made up of both real and fake face images is chosen to enable thorough inspection. Extensive experiments are conducted to evaluate the performance of the proposed models and compare them with the selected pre-trained models. The evaluation metrics employed include Accuracy, Precision, Recall, F1-score, Area Under the Receiver Operating Characteristic Curve (AUC-ROC), Receiver Operating Characteristic (ROC), and Confusion Matrix. These metrics provide a comprehensive analysis of the model's capability to accurately classify fake and real faces. The results demonstrate that Model 3 achieves highly promising performance in discriminating between fake and real faces. The comparison with pre-trained models reveals the superiority of Model 3 in terms of testing results, discrimination ability on new data, training, and validation performances. This study contributes to the field of fake vs. real face discrimination. The findings hold significant implications for applications in areas such as forensic analysis and social media content moderation.

Keywords: Classification Threshold · CNN · Evaluation Metrics · Fake Face Detection · GAN

1 Introduction

The increasing prevalence of manipulated or fake images and the sophistication of digital manipulation techniques, such as Generative Adversarial Networks (*GANs*) pose significant risks to privacy, security, and the authenticity of digital content. Additionally, the widespread availability of manipulated visual content threatens the authenticity of online information and compromises the security of

T. G. Debelee et al. (Eds.): PanAfriConAI 2023, CCIS 2069, pp. 219–241, 2024.
https://doi.org/10.1007/978-3-031-57639-3_10

facial recognition systems. This research paper aims to implement, train, and test different deep learning algorithms, specifically Convolutional Neural Networks (*CNN*s) with different hyperparameters and architectures. Additionally, conduct a thorough comparison between the implemented models and some pre-trained models like ResNet-50, Inception-ResNet-V2, DenseNet-121, and EfficientNet-B2 to reduce the generalizable error.

1.1   Generation of Fake Images Using Generative Adversarial Networks

Many algorithms are now being used to generate fake images, such as Generative Adversarial Networks (*GAN*s), which are a class of machine learning models that consist of two neural networks, namely the generator and the discriminator. In the training process, it involves training these two neural networks in a competitive setting. This training process is called Adversarial Training; it includes training a generator network and a discriminator network simultaneously, as illustrated in Fig. 1.

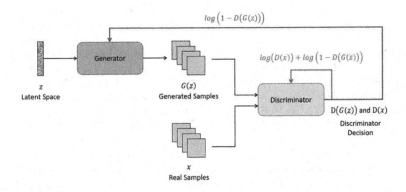

Fig. 1. Typical Generative Adversarial Networks (*GAN*) architecture [18].

Additionally, *GAN*s create artificial data that mimics real data through adversarial training. The creation of *GAN*s has allowed computers to produce realistic face images that are capable of fooling people [8]. Several well-known *GAN*s, such *StyleGAN*[1] and *PGGAN*[2], have been utilized to produce realistic fake face images. Recent times have witnessed notable progress in the creation of *GAN*s, or Generative Adversarial Networks, which are used to generate realistic facial images. Furthermore, fake faces observed in real-world situations come from multiple unknown sources, specifically distinct *GAN*s. The work becomes more hard due to the unanticipated picture distortions that these faces may

---

[1] *StyleGAN*: Style Generative Adversarial Network.
[2] *PGGAN*: Progressive Growing Generative Adversarial Network.

experience, including down-sampling, blur, noise, and JPEG compression [8]. As a result, even for human viewers, it has becoming harder to discriminate between real and computer-generated facial images [1].

2   Related Work

GANs have become a very useful tool in a wide range of computer vision applications quite quickly. In general, the way a *GAN* works involves two networks being trained in competition with each other. The 'generator' network is trained to produce artificial images that are almost identical to a dataset of real photos that is supplied. Meanwhile, the 'discriminator' is trained to discriminate correctly between images generated by the generator and those that are actual [11]. Advanced Generative Adversarial Network (*GAN*) models, such as *PGGAN* (see footnote 2) [5] and *StyleGAN* (see footnote 1) [6], can generate resolutions of up to (1024 × 1024) pixels that are of excellent quality. These photos are realistic enough to fool human beings. Each *GAN* leaves unique fingerprints on photos, as demonstrated by Marra et al. [9]. They suggested using raw pixels and traditional forensics features taken from actual and fake images to train to identify the source of these images. Yet, the approach is incapable of extrapolating its ability to identify fake faces generated by *GAN* models that were not included in the training dataset. Data augmentation was attempted by Xuan et al. [20] to improve the generalization process; however, further improvements are limited by the detection algorithm's constraints. Using global image texture representations, Liu et al. [8] have proposed a new architecture called Gram-Net that enables robust fake image detection. Additionally, it can recognize fake face images rather well and generalizes much better when detecting fake faces from *GAN* models not observed during the training phase. Using the co-occurrence matrix as a feature is recommended by Nataraj et al. [12], who show better performance than classifiers trained on raw pixel data from *CycleGAN*. According to McCloskey and Albright [11], there are certain color cue artifacts in images produced by *GANs* that are linked to the normalization layers. Consequently, the detecting procedure can make use of these artifacts. The absence of global limitations leads to a divergence in the facial arrangement between synthetic and actual photographs, as noted by Yang et al. [22]. Thus, the locations of facial landmark points alone can be used to train a Support Vector Machine. Wang et al. [19] introduced a fake detection method based on neuron coverage. However, a drawback is that the algorithm is time-intensive and challenging to implement in real systems. With the use of incremental learning, Marra et al. [10] were able to identify fake photos. However, the effectiveness of this strategy depends on the availability of several *GAN* models during training. Some of these approaches, including those addressing generalizability issues with various *GAN* models, can be both time-consuming and impractical, especially when resorting to training with multiple models.

3   Methods

3.1   Convolutional Neural Networks

Convolutional Neural Network, also known as *CNN*, is a type of deep learning algorithm that is primarily used for analyzing visual data such as videos and images. It has emerged as a prominent neural network in deep learning, playing a significant role in computer vision. Applications of *CNN* have enabled advances in facial recognition, self-driving cars, self-service supermarkets, and intelligent medical treatments-all of which were previously thought to be unattainable [7]. They are designed to extract relevant features from the input data through a process called convolution. *CNN*s are constructed with convolutional layers, pooling layers, and fully connected layers, which, when combined, constitute the architecture of a *CNN* [13]. In Fig. 2, a simplified *CNN* structure for MNIST classification is displayed.

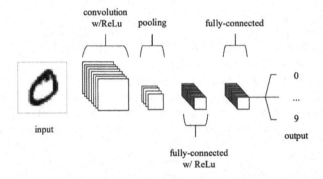

Fig. 2. A simplified *CNN* architecture for MNIST classification, comprised of just five layers [13].

The process of optimizing parameters such as kernels is called training, which is performed so as to minimize the difference between outputs and ground truth labels through an optimization algorithm called backpropagation and gradient descent, among others [21].

Loss Function: The binary cross entropy loss function, also known as log loss or logistic loss, is a commonly used loss function in binary classification tasks. It measures the dissimilarity between the predicted probabilities and the true binary labels of a classification problem. Binary cross entropy compares each of the predicted probabilities to the actual class output, which can be either 0 or 1. It then calculates the score that penalizes the probabilities based on the distance from the expected value. That means how close or far from the actual value. In the case of binary classification, where there are two classes (usually denoted as 0 and 1), the binary cross entropy loss function is defined as:

$$L(y, \hat{y}) = -(y \log(\hat{y}) + (1 - y) \log(1 - \hat{y}))$$

where y is the true binary label (either 0 or 1), and $\hat{y}$ is the predicted probability of the positive class, ranging from 0 to 1.

Swish Activation Function: A non-linear activation function that combines the linearity of the identity function for positive inputs with the non-linearity of the sigmoid function for negative inputs. It has been found to perform well in deep neural networks, providing better training performance and improved accuracy compared to other activation functions like $ReLU^3$ (Rectified Linear Unit).

The Swish activation function is defined as:

$$f(x) = x \cdot \sigma(\beta x)$$

where x is the input to the activation function, σ represents the sigmoid function, which is the following function: $\sigma(z) = (1 + \exp(-z))^{-1}$, and β is a trainable parameter that controls the slope of the function. It can be either a constant or a parameter that can be adjusted during training. Figure 3 illustrates the Swish function across various β values. When β is set to 1, Swish is essentially the Sigmoid-weighted Linear Unit (SiL) introduced by Elfwing et al. [2] for reinforcement learning. In the case of β equal to 0, Swish transforms into the scaled linear function $f(x) = \frac{x}{2}$. As β approaches infinity, the sigmoid component tends toward a 0-1 function, making Swish resemble the $ReLU$ (see footnote 3) function. This suggests that Swish can be seen as a smooth function that nonlinearly interpolates between the linear function and the $ReLU$ (see footnote 3) function. The extent of this interpolation can be controlled by the model when β is configured as a trainable parameter [14].

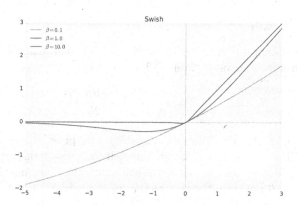

Fig. 3. The Swish activation function [14].

---

3 $ReLU$: Rectified Linear Unit.

Dropout: A regularization technique commonly used in deep learning neural networks. It helps prevent over-fitting by reducing the inter-dependencies between neurons or feature maps during training. In dropout, randomly selected neurons or features are ignored or "dropped out" during forward and backward passes, meaning their contributions to the activation of downstream neurons and gradients are temporarily removed (Fig. 4). The key idea is to randomly drop units (along with their connections) from the neural [15]. During training, each neuron in a layer has a probability (typically between 0.2 and 0.5) of being dropped out. This probability is a hyper-parameter that determines the amount of dropout applied. The dropped-out neurons are randomly selected for each training sample and each training iteration.

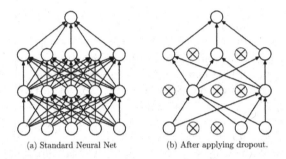

(a) Standard Neural Net   (b) After applying dropout.

Fig. 4. Dropout Neural Net Model. **Left**: A standard neural net with 2 hidden layers. **Right**: An example of a thinned net produced by applying dropout to the network on the left. Crossed units have been dropped [15].

3.2 Pixel Re-scaling

This process is done by dividing each pixel value in each image by 255 so that all pixels are in a smaller range from [0,255] to [0,1].

$$RescaledPixelValue = \frac{OriginalPixelValue}{255}$$

Pixel re-scaling provides numerical stability as pixel values are typically smaller and fall within a well-defined range. It can also enhance the model's ability to generalize to unseen data. Additionally, the learning algorithm can converge faster and navigate the parameter space more effectively.

3.3 140k Real and Fake Faces Dataset

This dataset[4] was used in all of the conducted experiments (proposed and pre-trained). According to the name, this dataset consists of all 70k real faces and

----

[4] https://www.kaggle.com/datasets/xhlulu/140k-real-and-fake-faces.

70k fake faces generated by *StyleGAN* (see footnote 1). In this dataset, all the images are of size 256px. All the images are in RGB format. The dataset is split into three separate dataset folders: the training dataset, the validation dataset, and the test dataset. The training dataset consists of 100,000 face images that are divided evenly into 50,000 real face images and 50,000 fake face images. Both the validation and test datasets are of size 20,000. Each of these two datasets is also split evenly (10,000 real face images and 10,000 fake face images).

3.4 Training Details

Some experiments were trained on a subset of the training dataset from the **140k Real and Fake Faces Dataset**, while the others utilized the entire training dataset. The GPU that was used for training was a GPU T4 x2. It includes two NVIDIA T4 GPUs and works well with image classification problems. The training dataset is processed in batches during training. For example, with a dataset of 100,000 examples and a batch size of 64, there are 1562 batches per epoch. The training dataset is shuffled before training. The loss function used was the Binary Cross Entropy loss function. The optimizer used during training was the **Adam**[5] optimizer with a learning rate of 0.0001. All images in the training dataset were resized to $224 \times 224 \times 3$. The last layer's activation function is the sigmoid function. The selected batch size is 64 for the training dataset. The model's performance on the training dataset is reflected in the accuracy reported for each epoch. In each epoch, it displays the percentage of training data examples that were correctly classified. During training, it is determined for each batch by contrasting the model's predictions with the batch's actual labels. The ratio of examples in the batch that was correctly predicted to all of the examples in the batch is used to calculate accuracy.

$$Training\ Accuracy = \frac{Number\ of\ correctly\ classified\ training\ samples}{Total\ number\ of\ training\ samples} \quad (1)$$

3.5 Validation Details

At the end of each epoch during training, the whole validation dataset is tested in batches, and the validation loss and accuracy are computed. The selected batch size was 64 for the validation dataset. In our case, the validation dataset consists of 20,000 images, split evenly into 10,000 real face images and 10,000 fake face images. Therefore, the validation batches would be $\frac{20,000}{64}$ batches, which is 312 batches in each epoch. The validation dataset is always shuffled. All images in the validation dataset were resized to $224 \times 224 \times 3$. The purpose of the validation dataset is to demonstrate how effectively the model generalizes to new data. For validation accuracy, it measures the percentage of correctly classified examples across the entire validation dataset. In Eq. 2, we can observe the formula for calculating the validation accuracy.

---

[5] *Adam*: Adaptive Moment Estimation.

$$Validation\ Accuracy = \frac{Number\ of\ correctly\ classified\ validation\ samples}{Total\ number\ of\ validation\ samples} \quad (2)$$

3.6 Testing Details and Model Evaluation

The test dataset also consists of 20,000 samples. The dataset is split evenly into 10,000 real face images and 10,000 fake face images. No shuffling is done in the test dataset. In the test dataset, all of the 20,000 samples are different from the samples that exist in both the training and validation datasets. The batch size chosen for the test dataset is 1. This was done to ensure that each test sample is evaluated individually and provides memory efficiency. All images in the test dataset were resized to $224 \times 224 \times 3$. After the training process is done, we then analyze any model on the whole test dataset to see its overall performance. The test accuracy is calculated by the following formula:

$$Test\ Accuracy = \frac{Number\ of\ correctly\ classified\ test\ samples}{Total\ number\ of\ test\ samples} \quad (3)$$

After calculating the test accuracy for each model, we calculate the F1 Score, Precision and Recall, Receiver Operating Characteristic curve with Area Under the Curve (AUC-ROC), Average Precision (AP), and Weighted and Macro Averages. In addition to plotting the Confusion Matrix and ROC Curve (Figs. 9, 11 and 13).

3.7 Proposed Models

Some common properties between all three proposed models are that the main activation function was chosen to be the Swish activation function and was utilized after every convolution with $\beta = 1$. Also, L2 Regularization was established in the second fully connected layer in all three models with a kernel regularizer that has a factor of 0.001. **Adam** (see footnote 5) optimizer was used with a learning rate of 0.0001, and Binary Cross Entropy was the loss function as mentioned in Sect. 3.4. All of the validation and test datasets were utilized as mentioned in Sects. 3.5 and 3.6. Max-pooling was applied in all the pooling layers. Pixel re-scaling was performed on all the images in the training, validation, and test datasets.

1. **Model 1:** The model operates on a subset of the training dataset from the **140k Real and Fake Faces** dataset during the training process. The subset consists of only 40,000 images that are divided evenly into 20,000 real face images and 20,000 fake face images. The model was trained for 12 epochs. The first model comprises:
   - Three Convolutional Layers with a kernel size of 3×3.
   - Three Pooling Layers with a pool window of size 2×2.
   - Dropout was applied after every convolution, after the last pooling layer, and in the second fully connected layer with factors 0.25 and 0.5.

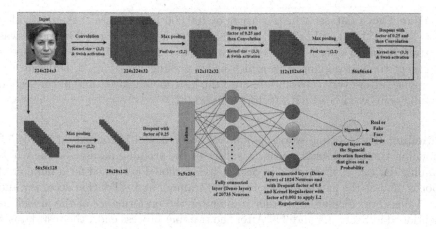

Fig. 5. Structure of Model 1.

2. **Model 2:** This model also operates on a subset of the training dataset from the **140k Real and Fake Faces** dataset; however, the subset consists of only 50,000 images that are divided evenly into 25,000 real face images and 25,000 fake face images. The model was trained for 13 epochs. The second model comprises:
   - Four Convolutional Layers with a kernel size of 3×3.
   - Four Pooling Layers. The pool size was 2×2 in all of the pooling layers except the last pooling layer which had a pooling window of size 3×3 to reduce spatial dimensionality of the output.
   - Dropout was also applied after every convolution, after the last pooling layer, and in the second fully connected layer, with a factor of only 0.25.
3. **Model 3:** This model has the same structure as the second model, however, this model operates on the whole training dataset from the **140k Real and**

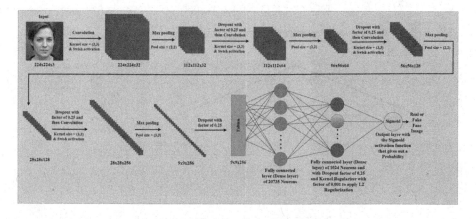

Fig. 6. Structure of Model 2.

Fake Faces dataset which consists of 100,000 face images that are divided evenly to 50,000 real face images and 50,000 fake face images. The model was trained for 15 epochs. The pool size was 2×2 in all of the pooling layers except the last pooling layer which had a pooling size of 3×3. Dropout was applied after every convolution, after the last pooling layer, and in the second fully connected layer with factors 0.1 and 0.2.

Pipeline: The pipeline works as follows: every model is trained on its' corresponding training dataset, either a subset of the training dataset or the whole training dataset as mentioned in Sect. 3.4. The model is trained for a number of epochs, and at the end of every epoch, the training and validation accuracy and loss values are calculated in order to monitor the performance of the model, as mentioned in Sects. 3.4 and 3.5. After the training process ends, the test dataset is fed into the model in order to calculate the test accuracy and other evaluation metrics, as stated in Sect. 3.6.

3.8 Pre-trained Experiments

In this part, pre-trained models are also trained and compared with the three proposed models above. The pre-trained models are Inception-ResNet-V2, ResNet-50, DenseNet-121, and EfficientNet-B2. These models had common properties, such as operating on the whole training dataset from the **140k Real and Fake Faces** dataset. Once more, all of the validation and test datasets were used. *Adam* (see footnote 5) optimizer was used with a learning rate of 0.0001, and Binary Cross Entropy was the loss function as mentioned in Sect. 3.4. All of these models were trained for 15 epochs. A Global Average Pooling layer was added after the output of every pre-trained model to reduce the spatial dimensions of the data before feeding it into the fully connected layer (Dense layer). Pixel

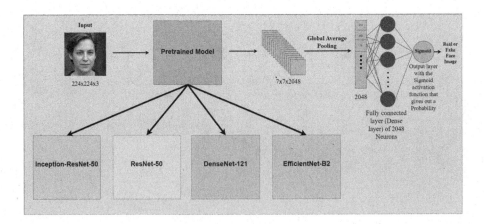

Fig. 7. Structure of the pre-trained models.

re-scaling was performed on all the images in the training, validation, and test datasets.

These models follow the pipeline mentioned in Sect. 3.7, however, all of these models use the whole training dataset, validation dataset, and test dataset from the **140k Real and Fake Faces** dataset. From Fig. 7, we can see that all 4 pre-trained models have the same architecture from the outside; however, these pre-trained models are different from each other in the inside architecture. Inception-ResNet-50 is a deep convolutional network architecture that includes components from the two well-known models, ResNet and Inception. Szegedy et al. [16] proposed this model architecture using the advantages of both the ResNet and Inception architectures. ResNet-50 is a convolutional neural network architecture. By incorporating shortcut connections, which enable the network to bypass one or more levels. It was introduced as part of the 'Deep Residual Learning' paper by He et al. (2015) [3]. DenseNet-121 is a particular variation of the DenseNet architecture. It was introduced in the 'Densely Connected Convolutional Networks' study by Huang et al. (2017) [4]. EfficientNet-B2 is a particular variation of the EfficientNet architecture. These EfficientNet architectures were introduced by Tan and Le [17].

4   Results and Discussions

4.1   Results of Model 1

Lets first examine the results of the Model 1 mentioned in Sect. 1.

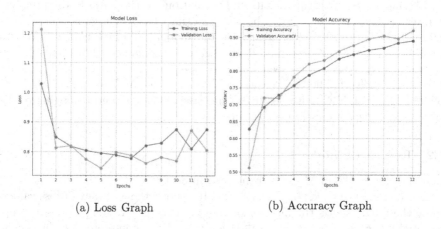

(a) Loss Graph   (b) Accuracy Graph

Fig. 8. Loss and Accuracy Graphs for Model 1 During the 12 Epochs of Training.

As seen in Figs. 8a and 8b, the x-axis represents the number of epochs during training, and the y-axis represents the loss or accuracy values. The blue curve represents training values, and the orange curve represents validation values. The training accuracy is calculated using Formula 1, and the validation accuracy is calculated using Formula 2. Let's analyze both graphs. There is a zigzag pattern noticed in the training and validation loss curves. This can be due to using a subset of the training dataset, so the model was not trained on a relatively good and wide dataset. Another factor could be the relatively large number of parameters, which was around 102,855,745. Therefore, the model had difficulty converging to the optimal point. The training and validation accuracy curves are smoother and show a better growth trajectory throughout the training process.

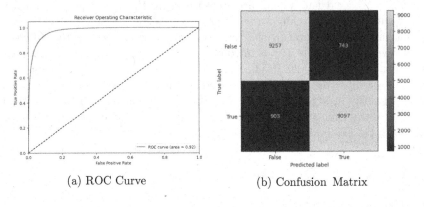

(a) ROC Curve   (b) Confusion Matrix

Fig. 9. ROC Curve and Confusion Matrix for Model 1.

Moving on to the testing results and evaluation metrics, after feeding the test dataset to the first model, it can be observed that the ROC area is estimated at 0.92, as seen in Fig. 9a. This means that our model has a high ability to distinguish between real (positive) and fake (negative) face images. For the confusion matrix in Fig. 9b, it can be seen that there were a lot of misclassified samples. This is due to having 903 false negatives (misclassified as fake and belonging to the real class) and 743 false positives (misclassified as real and belonging to the fake class). These numbers can be improved.

4.2   Results of Model 2

Here are the results of Model 2 mentioned in Sect. 2.

Once more, the y-axis in Figs. 10b and 10a shows the accuracy or loss values, while the x-axis shows the number of training epochs. Training values are represented by the blue curve, and validation values are represented by the orange curve. Formulas 1 and 2 are used to compute the training and validation accuracy, respectively. For Model 2, the training process had improved over Model 1. The training loss and accuracy curves are smoother and do not have any instabilities. This is due to having a reduced number of parameters, around 21,624,129. Another reason could be due to the use of a larger subset for training. The validation loss and accuracy still had some instabilities. The zigzag pattern was

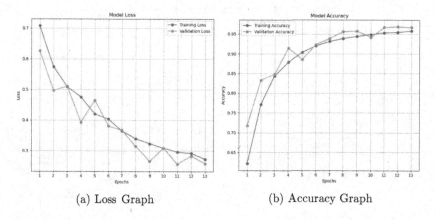

(a) Loss Graph   (b) Accuracy Graph

Fig. 10. Loss and Accuracy Graphs for Model 2 During the 13 Epochs of Training.

still present in the validation curves. In consequence, Model 2 did not generalize perfectly to new data.

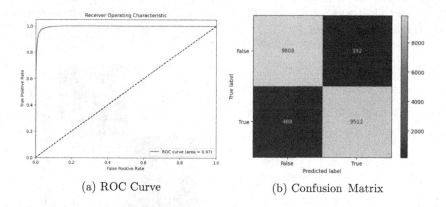

(a) ROC Curve   (b) Confusion Matrix

Fig. 11. ROC Curve and Confusion Matrix for Model 2.

After feeding the test dataset to Model 2, it can be observed that the ROC area is estimated at 0.97, as seen in Fig. 11a. Having examined the confusion matrix in Fig. 11b, it is noticed that the number of misclassifications (false positives and false negatives) is reduced compared to the Model 1. This means this model has better generalizability performance. Another note here is that this model misclassifies more in the real class than in the fake class (488 false negatives vs. 192 false positives). This leads to some sort of biased classification behaviour.

4.3   Results of Model 3

Here are the results of Model 3 mentioned in Sect. 3.

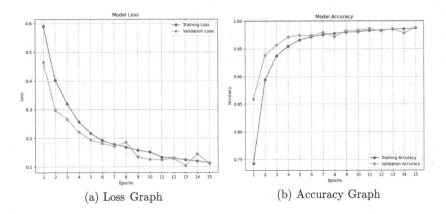

(a) Loss Graph   (b) Accuracy Graph

Fig. 12. Loss and Accuracy Graphs for Model 3 During the 15 Epochs of Training.

From Figs. 12a and 12b, Model 3 had the same number of parameters as the second model (21,624,129) and was trained on the whole training dataset. As a result, it gave the smoothest training and validation loss and accuracy curves. The zigzag effect was not present. Thus, this model had an obvious improvement over the first two models.

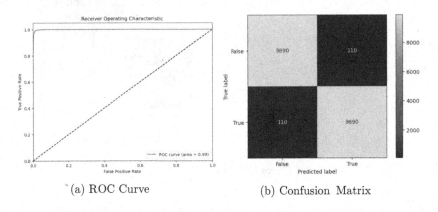

(a) ROC Curve   (b) Confusion Matrix

Fig. 13. ROC Curve and Confusion Matrix for Model 3.

Moving on to the testing results and evaluation metrics, after feeding the test dataset to Model 3, the ROC area is estimated at 0.99, as seen in Fig. 13a. This implies that Model 3 has so far had the highest ability to distinguish between real (positive) and fake (negative) face images. After observing the confusion matrix, it is noticed that the number of misclassifications is reduced even more compared to the first two models. On top of that, this model observes a symmetrical classification behaviour (unbiased), which can be seen in Fig. 13b, as the number of false positives is equal to the number of false negatives and the number of true positives is equal to the number of true negatives (110 and 9890). This means this model has given the best results so far.

4.4   Evaluation Metrics for the Three Proposed Models

The evaluation metrics are obtained by feeding the whole test data set into every model (Model 1, Model 2, and Model 3) after the training process, as mentioned in Sect. 3.6.

Table 1. Evaluation Metric Results for The Three Proposed Models.

Model	Metric	Class 0	Class 1	Macro Avg	Weighted Avg
Model 1	Precision	0.91	0.92	0.92	0.92
	Recall	0.93	0.91	0.92	0.92
	F1-Score	0.92	0.92	0.92	0.92
	Support	10,000	10,000	20,000	20,000
	ROC AUC Score			0.9759	
	AP Score			0.9728	
	Test Accuracy			0.9177	
Model 2	Precision	0.95	0.98	0.97	0.97
	Recall	0.98	0.95	0.97	0.97
	F1-Score	0.97	0.97	0.97	0.97
	Support	10,000	10,000	20,000	20,000
	ROC AUC Score			0.9957	
	AP Score			0.9953	
	Test Accuracy			0.9660	
Model 3	Precision	0.99	0.99	0.99	0.99
	Recall	0.99	0.99	0.99	0.99
	F1-Score	0.99	0.99	0.99	0.99
	Support	10,000	10,000	20,000	20,000
	ROC AUC Score			0.9993	
	AP Score			0.9992	
	Test Accuracy			0.9890	

From Table 1, it can be deduced that the best model in terms of results is the third model. Furthermore, **Support** indicates to the number of test samples used, class 0 is the fake class and class 1 is the real class. Test accuracy is calculated using Formula 3.

4.5   Results of Pre-trained Models

Here are the results of the pre-trained models, all of which were trained for 15 epochs and were trained on the whole training dataset as mentioned in Sect. 3.8.

These results are then compared to the results of our proposed models (Model 1, Model 2, and Model 3) obtained in Sects. 4.1, 4.2, 4.3, and 4.4. Similarly, the y-axis in Figs. 14, 15, 16, and 17, shows the accuracy or loss values, while the x-axis indicates the total number of training epochs. Validation values are represented by the orange curve, and training values are represented by the blue curve. Formula 1 is used to compute the training accuracy, and Formula 2 is used to calculate the validation accuracy.

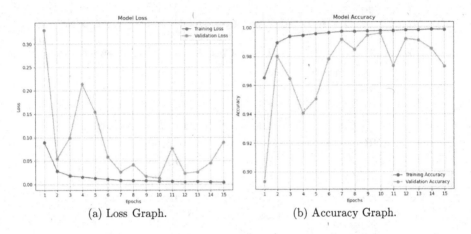

(a) Loss Graph.   (b) Accuracy Graph.

Fig. 14. Loss and Accuracy Graphs of Inception-ResNet-50 During the 15 Epochs.

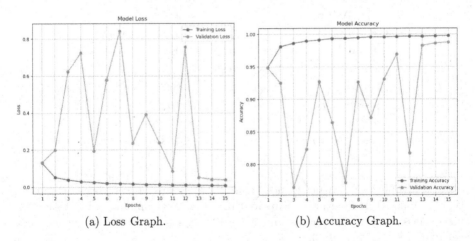

(a) Loss Graph.   (b) Accuracy Graph.

Fig. 15. Loss and Accuracy Graphs of ResNet-50 During the 15 Epochs.

It can be observed that all the pre-trained models had smooth training loss and accuracy curves. The training curves did not suffer from any oscillations;

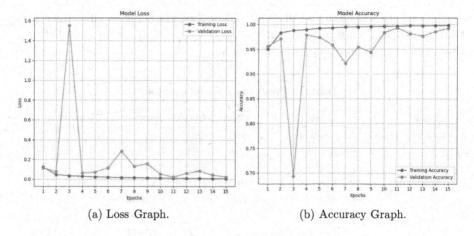

(a) Loss Graph.   (b) Accuracy Graph.

Fig. 16. Loss and Accuracy Graphs of DenseNet-121 During the 15 Epochs.

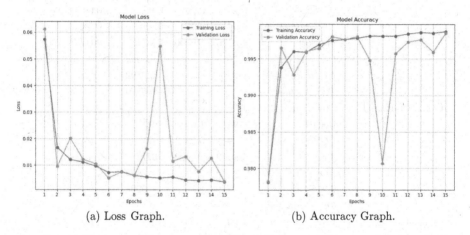

(a) Loss Graph.   (b) Accuracy Graph.

Fig. 17. Loss and Accuracy Graphs of EfficientNet-B2 During the 15 Epochs.

hence, they did not have a hard time converging to the optimal point and were considered better than the training curves seen in the three proposed models in Sects. 4.1, 4.2, and 4.3. The problem lied in the validation curves. Most pre-trained models were over-fitting, especially Inception-ResNet-V2 and ResNet-50, where the validation accuracy values were less than the training accuracy values. Also, the validation loss values are greater than the training loss values. This behaviour is noticed in Figs. 14a, 14b, 15a, and 15b. Therefore, most of these models had a harder time classifying unseen data, hence suffering from over-fitting. This means that our proposed models (Model 1, Model 2, and Model 3) had better loss and accuracy validation curves. This can be due to the fact that the three proposed models did not suffer from over-fitting; however, Model 1 Sect. 4.1 and Model 2 Sect. 4.2 did suffer from some oscillations, but these oscillations are considered small compared to the ones found in all of these pre-trained models. As a result, our three models had a better compromise between training and generalization performances.

Table 2. Pre-trained Model Results.

Model	Metric	Class 0	Class 1	Macro Avg	Weighted Avg
Inception-ResNet-V2	Precision	0.95	0.98	0.97	0.97
	Recall	0.98	0.95	0.97	0.97
	F1-Score	0.97	0.97	0.97	0.97
	Support	10,000	10,000	20,000	20,000
	ROC AUC Score			0.9998	
	AP Score			0.9997	
	Test Accuracy			0.9714	
ResNet-50	Precision	0.99	0.98	0.99	0.99
	Recall	0.98	0.99	0.99	0.99
	F1-Score	0.99	0.99	0.99	0.99
	Support	10,000	10,000	20,000	20,000
	ROC AUC Score			0.9992	
	AP Score			0.9992	
	Test Accuracy			0.9879	
DenseNet-121	Precision	1.00	0.98	0.99	0.99
	Recall	0.98	1.00	0.99	0.99
	F1-Score	0.99	0.99	0.99	0.99
	Support	10,000	10,000	20,000	20,000
	ROC AUC Score			0.9998	
	AP Score			0.9998	
	Test Accuracy			0.9914	
EfficientNet-B2	Precision	1.00	1.00	1.00	1.00
	Recall	1.00	1.00	1.00	1.00
	F1-Score	1.00	1.00	1.00	1.00
	Support	10,000	10,000	20,000	20,000
	ROC AUC Score			0.9999	
	AP Score			0.9999	
	Test Accuracy			0.9984	

Upon providing the test dataset as input to the pre-trained models, it was found that the ROC Curve area values ranged from 0.97 to 1, as seen in Table 4. This implies that all pre-trained models have an outstanding ability to distinguish between real (positive) and fake (negative) face images. Additionally, the ROC Area values in the pre-trained models are considered much better than the ones obtained from our three proposed models (Model 1, Model 2, and Model 3) in Figs. 9a, 11a, and 13a. For the confusion matrices, it is evident from Table 3

Table 3. Comparison of TP, FP, TN, FN for Four Models.

Model	True Positive	False Positive	True Negative	False Negative
Inception-ResNet-V2	9999	570	9430	1
ResNet-50	9915	158	9842	85
DenseNet-121	9983	156	9844	17
EfficientNet-B2	9991	23	9977	9

Table 4. Comparison of ROC Area for Four Pre-trained Models.

Model	ROC Area
Inception-ResNet-V2	0.97
ResNet-50	0.99
DenseNet-121	0.99
EfficientNet-B2	1

that most pre-trained models have a higher false positive rate than a false negative rate. This means that most of these pre-trained models misclassify fake face images as real face images. This can be an issue when these models are introduced to new unseen images from a different distribution, especially if these images are fake face images. Therefore, this is considered to be a biased classification behaviour for all of these models. As a result, this contradicts the fact that these models have good generalization performance and an outstanding ability to discriminate between real and fake face images. Consequently, Model 3, EfficientNet-B2, and DenseNet-121 have obtained the best confusion matrices, as the number of misclassifications in both the fake and real classes is low in all of them. Moving on to the evaluation metrics for pre-trained models. After feeding the test dataset, It can be observed from Table 2 that all of the pre-trained models obtained outstanding results, EfficientNet-B2 in particular. When comparing these evaluation metric results with the ones obtained from our three proposed models in Sect. 4.4, it is easy to say that these pre-trained models provided better evaluation metrics. Previously, it was noticed that these pre-trained models obtained a biased classification behavior, and some suffered from over-fitting in the training process; however, they gave outstanding results. Therefore, we need to perform a validation test to validate these results.

4.6 Validation Test

The purpose of the test was to make sure that the results obtained above for every model held true. This test involves feeding five fake face images to each model and checking if each model can correctly classify the images, including unseen data from a different distribution. For every model, it predicts whether the image is fake or real using a chosen classification threshold value, symbolized as γ. The classification function $f_\gamma(x)$ can be defined as follows:

$$f_\gamma(x) = \begin{cases} \text{Fake,} & \text{if } x < \gamma, \\ \text{Real,} & \text{if } x \geq \gamma, \end{cases}$$

where x represents the prediction probability value and γ is the threshold value for classification. In our experiments, γ was chosen to be 0.5. As seen in Fig. 18, there are five fake face images of people with different expressions and different background light settings. These images were generated using *StyleGAN* (see footnote 1).

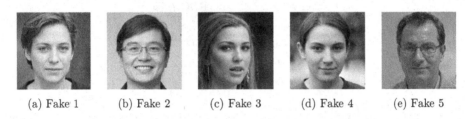

(a) Fake 1   (b) Fake 2   (c) Fake 3   (d) Fake 4   (e) Fake 5

Fig. 18. Fake Face Images (Fake Face Images Generated by https://thispersondoes notexist.com/).

Table 5. Validation Test Results.

Fake Elements						
Model	1	2	3	4	5	Accuracy
Model 1	4.733×10^{-9}	0.0004	0.0990	4.782×10^{-5}	0.0040	100.0%
Model 2	1.496×10^{-6}	0.0018	0.0008	3.169×10^{-5}	0.0020	100.0%
Model 3	7.115×10^{-6}	0.0130	0.0107	0.1280	0.0800	100.0%
Inception-ResNet-V2	0.9990	1.0000	0.9990	1.0000	0.9990	0.0%
ResNet-50	0.3510	0.9990	0.9980	0.9950	0.0010	40.0%
DenseNet-121	0.0020	0.1175	0.0087	0.9820	0.0140	80.0%
EfficientNet-B2	0.9990	0.9990	0.9990	0.9990	0.9990	0.0%

After performing the test, it can be seen from Table 5 that the results are surprising. The accuracy here is the percentage of correctly classified images across the five fake images. Our three proposed models (Model 1, Model 2, and Model 3) performed the best with accuracies equal to 100%. On the other side, the pre-trained model delivered from horrible to good performances, with accuracies ranging from 0% to 80%. Only DenseNet-121 performed considerably well. EfficientNet-B2 had an accuracy of 0%, although it had the best results compared to all models, which seems unexpected. In addition, the biased classification behavior can be observed further here. Therefore, these pre-trained models need further tuning. Another note is that the three proposed models can misclassify unseen fake face images; however, the misclassification rate of the three proposed models is lower than the misclassification rate of the pre-trained models. Therefore, a takeaway here is that to enhance the robustness of the three proposed models (Model 1, Model 2, and Model 3) and the pre-trained models, it is recommended to include fake face images generated from multiple *GANs* in the training dataset so that each model would be prepared for all new and different types of fake face images to be deployed in real-life applications.

5 Conclusion

5.1 Work Summary

In this paper, many experiments were performed to distinguish between fake and real face images. This was done by proposing three *CNN* model architectures and comparing them with four pre-trained models (Inception-ResNet-V2, ResNet50, DenseNet-121, and EfficientNet-B2) that are considered state of the art when it comes to image classification tasks. The comparison was done based on the training and validation results, loss and accuracy graphs, evaluation metrics, and testing results for every model. Lastly, we performed a validation test to validate the obtained results by having five fake face images that are different from the ones used for training, validation, and testing. All models had good training performance and testing results. It was observed that all pre-trained models obtained biased classification behavior as the misclassification rate was greater for the fake class compared to the real class. Most of the pre-trained models were over-fitting, such as Inception-ResNet-V2. The best-performing pre-trained model was DenseNet-121. Finally, Model 3 in Sect. 3 was considered the best model as it had the best compromise between training performance, generalization performance, and testing results.

5.2 Limitations

Some limitations were encountered during the implementation, such as the hardware limitation, as the provided GPUs used in training had a limited time to be utilized. As a result, the training of all models took longer than it should have been. Another limitation was the unavailability of fake samples from different *GANs*. The dataset included only fake samples generated by *StyleGAN* (see footnote 1).

5.3 Future Work

To extend this work, Ensembles can be used to create a stronger model. Another recommendation is to test with different hyperparameters, such as a different number of epochs, dropout factors, and the number of input image pixels. Lastly, to utilize a dataset that contains fake samples from different *GANs* such as $DCGAN^6$, *PGGAN* (see footnote 2).

---

[6] *DCGAN*: Deep Convolutional Generative Adversarial Network.

References

1. Dang, L.M., Hassan, S.I., Im, S., Lee, J., Lee, S., Moon, H.: Deep learning based computer generated face identification using convolutional neural network. Appl. Sci. **8**(12), 2610 (2018)
2. Elfwing, S., Uchibe, E., Doya, K.: Sigmoid-weighted linear units for neural network function approximation in reinforcement learning. Neural Netw. **107**, 3–11 (2018)
3. He, K., Zhang, X., Ren, S., Sun, J.: Deep residual learning for image recognition. In: 2016 IEEE Conference on Computer Vision and Pattern Recognition (CVPR), pp. 770–778 (2016). https://doi.org/10.1109/CVPR.2016.90
4. Huang, G., Liu, Z., Van Der Maaten, L., Weinberger, K.Q.: Densely connected convolutional networks. In: 2017 IEEE Conference on Computer Vision and Pattern Recognition (CVPR) (2017)
5. Karras, T., Aila, T., Laine, S., Lehtinen, J.: Progressive growing of GANs for improved quality, stability, and variation. arXiv preprint arXiv:1710.10196 (2017)
6. Karras, T., Laine, S., Aila, T.: A style-based generator architecture for generative adversarial networks. IEEE Trans. Pattern Anal. Mach. Intell. **43**(12), 4217–4228 (2021). https://doi.org/10.1109/TPAMI.2020.2970919
7. Li, Z., Liu, F., Yang, W., Peng, S., Zhou, J.: A survey of convolutional neural networks: analysis, applications, and prospects. IEEE Trans. Neural Netw. Learn. Syst. **33**(12), 6999–7019 (2022). https://doi.org/10.1109/TNNLS.2021.3084827
8. Liu, Z., Qi, X., Torr, P.H.: Global texture enhancement for fake face detection in the wild. In: 2020 IEEE/CVF Conference on Computer Vision and Pattern Recognition (CVPR), pp. 8057–8066 (2020). https://doi.org/10.1109/CVPR42600.2020.00808
9. Marra, F., Gragnaniello, D., Cozzolino, D., Verdoliva, L.: Detection of GAN-generated fake images over social networks. In: 2018 IEEE Conference on Multimedia Information Processing and Retrieval (MIPR), pp. 384–389 (2018). https://doi.org/10.1109/MIPR.2018.00084
10. Marra, F., Saltori, C., Boato, G., Verdoliva, L.: Incremental learning for the detection and classification of GAN-generated images. In: 2019 IEEE International Workshop on Information Forensics and Security (WIFS), pp. 1–6 (2019). https://doi.org/10.1109/WIFS47025.2019.9035099
11. McCloskey, S., Albright, M.: Detecting GAN-generated imagery using saturation cues. In: 2019 IEEE International Conference on Image Processing (ICIP), pp. 4584–4588 (2019). https://doi.org/10.1109/ICIP.2019.8803661
12. Nataraj, L., et al.: Detecting GAN-generated fake images using co-occurrence matrices. arXiv preprint arXiv:1903.06836 (2019)
13. O'Shea, K., Nash, R.: An introduction to convolutional neural networks. arXiv preprint arXiv:1511.08458 (2015)
14. Ramachandran, P., Zoph, B., Le, Q.V.: Searching for activation functions. arXiv preprint arXiv:1710.05941 (2017)
15. Srivastava, N., Hinton, G., Krizhevsky, A., Sutskever, I., Salakhutdinov, R.: Dropout: a simple way to prevent neural networks from overfitting. J. Mach. Learn. Res. **15**(1), 1929–1958 (2014)
16. Szegedy, C., Ioffe, S., Vanhoucke, V., Alemi, A.A.: Inception-v4, inception-resnet and the impact of residual connections on learning. In: Proceedings of the Thirty-First AAAI Conference on Artificial Intelligence, AAAI 2017, pp. 4278–4284. AAAI Press (2017)
17. Tan, M., Le, Q.: Efficientnet: rethinking model scaling for convolutional neural networks. In: International Conference on Machine Learning, pp. 6105–6114. PMLR (2019)

18. Vint, D., Anderson, M., Yang, Y., Ilioudis, C., Di Caterina, G., Clemente, C.: Automatic target recognition for low resolution foliage penetrating SAR images using CNNs and GANs. Remote Sens. **13**(4) (2021). https://doi.org/10.3390/rs13040596
19. Wang, R., et al.: Fakespotter: a simple yet robust baseline for spotting AI-synthesized fake faces. arXiv preprint arXiv:1909.06122 (2019)
20. Xuan, X., Peng, B., Wang, W., Dong, J.: On the generalization of GAN image forensics. In: Sun, Z., He, R., Feng, J., Shan, S., Guo, Z. (eds.) Chinese Conference on Biometric Recognition. LNCS, vol. 11818, pp. 134–141. Springer, Cham (2019). https://doi.org/10.1007/978-3-030-31456-9_15
21. Yamashita, R., Nishio, M., Do, R.K.G., Togashi, K.: Convolutional neural networks: an overview and application in radiology. Insights Imaging **9**, 611–629 (2018)
22. Yang, X., Li, Y., Qi, H., Lyu, S.: Exposing GAN-synthesized faces using landmark locations. In: Proceedings of the ACM Workshop on Information Hiding and Multimedia Security, IH&MMSec 2019, pp. 113–118. Association for Computing Machinery (2019). https://doi.org/10.1145/3335203.3335724

Hybrid of Ensemble Machine Learning and Nature-Inspired Algorithms for Divorce Prediction

Kalkidan A. Sahle[1,2]($\boxtimes$) and Abdulkerim M. Yibre[1]

[1] Department of Information Technology, Faculty of Computing, Bahir Dar Institute of Technology, Bahir Dar University, Bahir Dar, Ethiopia
abdulkerim.mohamed@bdu.edu.et
[2] Department of Information Technology, Faculty of Informatics, Hawassa Institute of Technology, Hawassa University, Hawassa, Ethiopia
kalkidanalayu@hu.edu.et

Abstract. Divorce is a global issue with profound emotional, psychological, and socio-economic consequences. In 2022, Addis Ababa witnessed 14,000 registered marriages but also recorded 1,623 divorces, while 2018 saw 1,923 divorces. Understanding the factors contributing to divorce is vital for prevention and support. Machine learning and AI play a critical role in predicting divorce, early marital distress detection, and personalized interventions. Their scalability aids in effective prevention strategies and targeted support. This research explores a Hybrid Approach AdaBoost, Gradient Boosting, Bagging, Stacking, XGBoost, and Random Forest with Jaya and Whale Optimization. The Hybrid Approach is chosen to synergize the strengths of ensemble learning and nature-inspired optimization algorithms. The goal is to enhance divorce prediction accuracy by leveraging ensemble models' robustness and optimization inspired by natural processes. To assess model performance, train-test splits, and k-fold Cross-Validation techniques are used, with metrics like accuracy, precision, recall, F1 score, and AUC-ROC(Area Under the Receiver Operating Characteristic Curve). AdaBoost stands out, achieving 97%, and 96% accuracy in Jaya and WOA hyperparameter optimizations, respectively. This research aligns with Sustainable Development Goals (SDGs) by promoting gender equality (SDG 5), identifying inequalities and offering targeted support (SDG 10), and fostering stable families and social cohesion (SDG 16). By leveraging AI for divorce prediction, this work contributes to a more sustainable and inclusive world, advancing gender equality, reducing inequalities, and peaceful societies.

Keywords: Divorce · Ensemble Machine Learning · Nature-inspired Optimization Algorithms · Sustainable Development Goals · XGBoost

1 Introduction

Divorce is a significant life event that involves the legal dissolution of a marriage or marital union. It represents the termination of a relationship that was intended to be lifelong, affecting not only the couple involved but also their families, children, and society as a whole. Divorce can have profound emotional, social, and financial consequences [2], making it a subject of extensive research and societal interest. Understanding the factors that contribute to divorce and developing effective prediction models can aid in providing support, counseling, and interventions to couples in distress, as well as inform policies and practices related to marriage and family dynamics.

Divorce rates have been increasing in many countries over the past few decades, leading to a growing interest in studying the factors associated with marital dissolution. In Ethiopia, marriage holds considerable importance because of its deep-rooted religious and cultural significance. However, divorce rates have experienced a significant upswing, particularly in urban areas. According to research conducted in Addis Ababa, in the year 2022, there were 14,000 registered marriages and 1,623 recorded divorces. Similarly, in 2018, there were 1,923 registered divorces. Various social, cultural, economic, and individual factors contribute to the complexity of divorce. Research has extensively investigated various demographic variables and psychological factors associated with divorce. Demographic variables such as age, education, income, and employment status are influential predictors of divorce [6,12]. For instance, studies have shown that marrying at a younger age and having lower levels of education and income are associated with higher divorce rates. Unemployment or unstable employment patterns have also been linked to an increased risk of divorce.

Psychological factors play a crucial role in marital stability. Communication patterns, conflict resolution skills, and relationship satisfaction levels have been studied to understand their impact on divorce [12]. Poor communication, frequent conflicts, and ineffective resolution of disputes have been identified as risk factors for marital dissolution. In contrast, couples who demonstrate effective communication, healthy conflict management strategies, and higher levels of relationship satisfaction are more likely to have stable marriages.

Predicting divorce is a complex task that requires understanding the underlying dynamics of marriages and identifying key factors that contribute to marital dissolution. Traditionally, divorce prediction has relied on subjective assessments by professionals such as therapists or counselors. However, the limitations of these approaches have led researchers to turn to data-driven methods and machine learning algorithms to develop predictive models. By leveraging large-scale datasets and advanced analytical techniques, researchers aim to identify patterns, risk factors, and early indicators that can improve divorce prediction accuracy.

While ensemble learning has shown promise in improving prediction accuracy, its application in divorce prediction research is still relatively limited. Further exploration of ensemble methods, such as stacking, bagging, and boosting, could enhance the performance and robustness of divorce prediction models [16].

To improve divorce prediction models, this research explores a hybrid approach that combines ensemble learning techniques and nature-inspired optimization algorithms. The application of ensemble learning techniques, such as AdaBoost, Gradient Boosting, Bagging, Stacking, XGBoost, and Random Forest, in divorce prediction research remains relatively limited. Further research is needed to comprehensively investigate and compare the performance of these ensemble methods for divorce prediction tasks [13, 16].

Additionally, nature-inspired optimization algorithms, including Jaya and Whale optimization [15], have demonstrated efficacy in optimizing hyperparameters and improving the performance of machine learning models. However, their application in the context of divorce prediction is still underexplored. Further research is needed to evaluate and compare the effectiveness of various bio-inspired algorithms, including ant colony optimization, bat algorithm, and gray wolf optimization algorithms by combining these algorithms with other feature selection methods, researchers can explore novel optimization techniques and feature selection strategies to further enhance the performance and accuracy of divorce prediction models [1].

2   Related Works

Research in the realm of divorce prediction utilizing machine learning and ensemble machine learning algorithms has made substantial progress over the years. Numerous studies have been conducted to explore predictive models that can effectively anticipate divorce outcomes. These works encompass a variety of approaches, from traditional statistical models to advanced machine learning techniques. Researchers have identified various predictive factors, including relationship dynamics, socio-demographic variables, and psychological indicators. Machine learning algorithms such as support vector machines and logistic regression have been employed in a previous studies to build accurate models for divorce prediction.

Al-Behadili & Ku-Mahamud [1] presented a hybrid technique that combines K-Nearest Neighbors (KNN) and Particle Swarm Optimization (PSO) algorithms for divorce classification. The proposed technique was evaluated using the 10-fold cross-validation method and compared with other state-of-the-art classifiers. The results showed that the proposed technique achieved the best classification accuracy of 99.41%, which is very close to perfect classification. The technique also outperformed other classifiers in terms of precision, recall, F-measure, and Kappa statistic measurement. The author(s) concluded that the proposed technique can effectively predict divorce classification and find the appropriate number of features that represent the divorce dataset. Overall, the paper presents a novel approach to divorce classification using machine learning techniques and provides promising results.

As indicated in reference [14], Ranjitha & Prabhu introduced a novel approach for divorce prediction utilizing a dataset sourced from the UCI Machine Learning Repository. The dataset comprises 170 instances, each containing 54 attributes, and has undergone preprocessing techniques such as binning and normalization. The Particle Swarm Optimization (PSO) algorithm is employed in the preprocessing phase. During the training phase, 136 instances are used, while the remaining 34 instances are reserved for testing. The proposed method achieves a remarkable accuracy of 99.67%, surpassing existing techniques in the field. These findings provide valuable guidance to professionals involved in marital counseling, allowing them to schedule appropriate sessions to address marital discord effectively. The paper suggests that the proposed method can be extended to larger datasets to further enhance prediction accuracy and be utilized for case formulation and intervention planning.

Sadiq Fareed et al. [17] proposed a novel research study on predicting divorce prospects using ensemble learning techniques. Perceptron, logistic regression, neural networks, and randomized forest are four automatic learning models that the authors introduced, along with three hybrid models based on voting criteria. The author(s) compared the accuracy of several machine learning classifiers for divorce case prediction and achieved an accuracy of 98% with the perceptron model. They also conducted exploratory data analysis to determine the major factors that cause divorce and used feature engineering and data normalization to create a predictable model. The proposed ensemble learning approach based on three machine learning techniques achieved an accuracy score of 100% for predicting divorce. The authors identified key indicators for divorce and the factors that are most significant when predicting divorce.

A study conducted by Simanjuntak et al. [20] on predicting divorce using the backpropagation artificial neural network algorithm. The study uses different feature selection techniques and optimizes parameters such as training cycles, learning rate, and momentum to improve accuracy. The results show that the Relief feature selection technique has the highest accuracy level of 99.41%. The study also compares the accuracy of the backpropagation algorithm with and without feature selection and finds that the algorithm with feature selection performs better with an accuracy of 98.82%. Overall, the study provides insights into predicting divorce and highlights the potential of using backpropagation neural networks for this purpose.

Shankhdhar et al. [19] employed a total of six classification techniques, namely logistics regression, K-nearest neighbors, support vector machine (SVM), kernel SVM, decision tree, and random forest. Upon analyzing the results of these classification algorithms, it was found that four out of the six algorithms (logistic regression, SVM, decision tree, and random forest) exhibited similar accuracy percentages. The remaining two algorithms, K-nearest neighbors and kernel SVM, achieved accuracy rates of 97.93% and 98.62% respectively. To identify the most important attributes, correlation-based feature selection was utilized in conjunction with the classification algorithms. Consequently, out of the initial 53 features, a subset of 7 features was deemed effective and considered for further analysis.

The Divorce Predictors Scale developed by Yöntem et al. [24] within the scope of Gottman couples therapy can predict divorce rates with a 98.23% accuracy. The most effective predictors of divorce were related to creating a common meaning and failed attempts to repair, love map, and negative conflict behaviors. The study concluded that the Divorce Predictors Scale can predict divorce and may be beneficial for ministries that have direct contact with families, counseling services staff working on family counseling and family therapies, and for the preparation of case formulation and intervention plans. It was also concluded that the divorce predictors in the Gottman couples therapy were confirmed in the Turkish sampling.

A study by Dagnew et al. [8], aimed to estimate the prevalence of divorce from the first union and its predictors among reproductive-age women in Ethiopia, Tilson and Larsen [21] looked at two factors that may have an impact on the risk of divorce in Ethiopia: early age of first marriage, and childlessness within the first marriage, A study by Asfaw and Alene [3] an attempt has been done to assess and identify the major variables that contribute to divorce in Ethiopia. Age of women, place of residence, religion of a woman, education level of a woman, religion of a woman, and number of living children are the major variables that influence divorce in Ethiopia using a binary logistic regression model.

Overall, the majority of the studies in the related works used a questionnaire datasets consisting of 54 questions, and they have been responded to by a total of 170 individuals, comprising 84 divorced individuals and 86 married couples [1,14, 17,19,20,24] and Some of the author(s) did not employ optimization algorithms [19,24]. Additionally, Most of the previous studies conducted in Ethiopia used a statistical analysis to determine the risk factors for divorce [3,8,21].

3   Methodology

In this study, an experimental research design is applied. The utilization of an experimental research design in divorce prediction research helps ensure internal validity, as it enables researchers to establish cause-and-effect relationships between variables and outcomes. The study employed a methodology that involved collecting data through a questionnaire, performing exploratory data analysis, and applying pre-processing techniques such as applying encoding techniques, normalization, filling missing values, and class balancing. The data set was then used for model development using ensemble algorithms, with both train-test split and cross-validation approaches. Performance evaluation was conducted using various performance metrics. Hyperparameter optimization of the ensemble algorithms was achieved using Jaya and Whale optimization algorithms. The best performing model, based on optimized hyperparameters, was selected. The research methodology for this study is examined in Fig. 1.

3.1   Data Collection

The data utilized in this study comprises responses from a questionnaire, which was gathered through a combination of online surveys administered via KoboToolbox and traditional printed questionnaires. A questionnaire dataset was designed based on ideas from domain experts to gather information from individuals. The dataset comprises responses from 553 couples after removing duplicates, represented as individual records, to a comprehensive questionnaire containing 64 questions and a class to determine the current marital status of couples (divorced or not). Among these couples, 65 have experienced divorce,

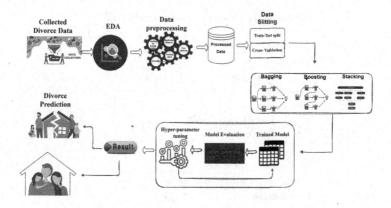

Fig. 1. Methodological framework diagram

while the remaining 488 are classified as Not Divorced. The survey consisted of questions that could be answered with "Yes", "No" or Neutral as well as questions that used a graded scale, including "Strongly Agree," "Agree," "Neutral," "Disagree," or "Strongly Disagree". There are some open-ended and close-ended questions as well. The dataset encompasses a wide range of variables covering demographic, affective, cognitive, economic, behavioral, health, and family factors that offer insights into the couples' characteristics and dynamics, such as age, gender, education level, affective experiences, cognitive traits, economic situations, behavioral patterns, health conditions, and family-related aspects like infertility or having a child from another partner. The data was gathered from three urban centers, namely Addis Ababa, Bahir Dar, and Hawassa. Table 1, presents an analysis of the divorce dataset (Table 2).

3.2   Exploratory Data Analysis

During the exploratory data analysis (EDA) phase, researchers employ diverse techniques to examine the dataset. Descriptive statistics, such as measures of central tendency (e.g., mean, median) and dispersion (e.g., standard deviation, range), are employed to provide concise summaries of the data [18]. Additionally, visualizations such as histograms, box plots, and scatter plots are utilized to graphically represent the distribution and relationships within the data [7]. These visualizations assist in identifying patterns, trends, and possible outliers within the dataset. In the bar chart presented in Fig. 2, "0" represents the Not Divorced class, and "1" represents the Divorced class. The bar plot visualizes the distribution of classes within a dataset. In this case, it represents the distribution of Divorced, and Not Divorced classes after balancing it using SMOTE to prevent the model from being biased toward the majority class.

Figure 3, showcases histogram charts representing the dataset. A histogram, a tool for visualizing data, resembles a bar chart where a range of outcomes is grouped into columns along the x-axis. The y-axis displays the frequency of

Table 1. Divorce dataset attributes

No	Questions by Domain Experts
1	Your age
2	Your spouse's age
3	Gender
4	Your ethnic origin
5	Your spouse's ethnic origin
6	Your occupation
7	Your spouse's occupation
8	Your educational status
9	Your spouse's educational status
10	Your religion
11	Your spouse's religion
12	We have miscommunication and not talking enough
13	We have behavioral incompatibility
14	We have special time together
15	we have similar view about how marriage should be
16	My spouse and I are not compatible sexually
17	There is a religious difference between us
18	My spouse and I have financial problems
19	I know how to take care for my sick spouse
20	I am unhappy with our marriage
21	I did not find marriage as I expected
22	My relatives interfere in our marriage
23	My spouse's relatives interfere in our marriage
24	I have another love relationship besides my spouse
25	My spouse has another love relationship besides me
26	I'm not the one who's guilty of what I am accused of
27	There is a wide age difference between us
28	Work consumes my time, leaving little for family
29	My spouse prioritizes work over family life
30	I suspect my spouse
31	My spouse suspects me
32	I have a jealous spirit
33	My spouse has a jealous spirit
34	We fulfill our responsibilities properly
35	My spouse is infertile
36	I am infertile
37	We frequently live apart for a long time due to work
38	My spouse has mental illness
39	I have mental illness
40	My spouse beats me
41	I beat my spouse
42	My spouse and I are insulting each other
43	My spouse is addicted
44	I am addicted
45	There is other people living in the same house with us

Table 2. Divorce dataset attributes (continued)

No	Questions by Domain Experts
46	My spouse feels inferior in the community
47	I feel inferior in the community
48	I know exactly what my spouse likes
49	My spouse knows exactly what I like
50	I feel aggressive when I argue with my spouse
51	My spouse feels aggressive when we argue
52	Our dreams with my spouse are similar and harmonious
53	Our home is not enough/not comfortable
54	My spouse is very stingy
55	I have child from another partner
56	My Spouse has child from another partner
57	My spouse have got physical injury
58	I have got physical injury
59	I know my spouse's friends and their social relationships
60	I know my spouse very well
61	I know my spouse's favorite food
62	We don't have time at home as partners
63	We are like two strangers who share the same home
64	We lived harmoniously together

numerical values occurring in the dataset columns, providing a clear depiction of the distribution of values and their occurrence rates. Overall, the collection of histograms provides a visual representation of the distribution and intensity of various factors contributing to divorce based on the survey responses.

The violin chart depicted in Fig. 4, serves as a visualization tool for exploring the reasons behind divorce within the dataset. This type of graph combines features of a kernel density plot and a box plot, offering a comprehensive view of data distribution and peaks. Its primary purpose is to illustrate the distribution of numerical data points within the dataset, providing insights into the patterns and concentration of factors contributing to divorce. The width of the "violins" reflects the probability density of the data, and the height indicates the frequency of responses at different values.

3.3 Data Preprocessing

Prior to utilizing the data for machine learning, several preprocessing techniques were implemented to ensure data quality and suitability for analysis. The preprocessing steps included handling missing values through mean and mode imputation, eliminating duplicate entries, and applying appropriate encoding techniques. For categorical variables, one-hot encoding was used for nominal columns, while label encoding was employed for ordinal columns. To address outliers, z-score standardization was utilized. Data normalization was performed using the min-max standardization technique to bring all features to a common

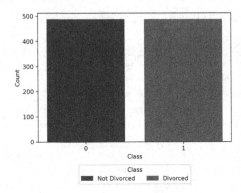

Fig. 2. Class Distribution in divorce dataset after balancing the data

scale. Lastly, class balancing was achieved using the Synthetic Minority Over-sampling Technique (SMOTE) [23]. These preprocessing techniques collectively prepared the data for machine learning, optimizing its suitability for subsequent analysis and modeling tasks.

3.4 Dataset Splitting

The proposed work utilized different ratios for the train-test split, including 70/30, 75/25, and 80/20. Additionally, the work also employed different values for the number of folds used in cross-validation, such as 5, 7, and 10. These variations in the experimental setup allowed for a more thorough evaluation of the model's performance and its ability to generalize to unseen data.

4   Proposed Ensemble Algorithms

4.1 Boosting

The work utilizes powerful and versatile boosting algorithms, including Ada-Boost, gradient boosting, and XGBoost, which are widely employed in machine learning. Ada-Boost assigns weights to training samples and iteratively trains weak learners, while gradient boosting sequentially corrects errors made by previous learners. XGBoost is an optimized framework that incorporates regularization and parallel processing. Recent studies highlight the effectiveness of these algorithms in various domains [5,9,10]. Ongoing research focuses on enhancing their performance and scalability.

4.2 Bagging

Bagging and Random Forest are ensemble learning techniques utilized in this work. Bagging trains multiple independent decision trees on bootstrap samples to enhance model stability, while Random Forest further improves performance by

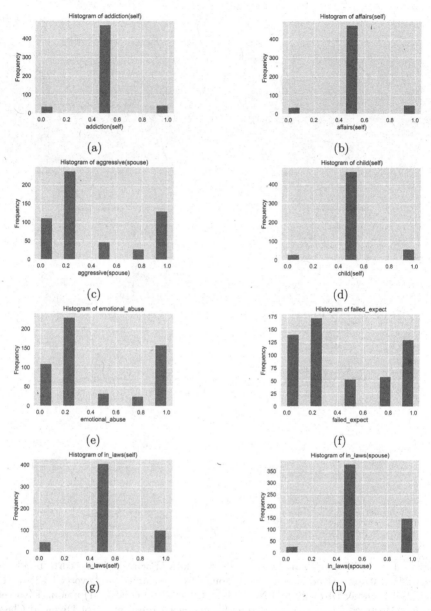

Fig. 3. The divorce histogram analysis of 8 prominent questions scale ranks analysis. Each subfigure corresponds to a specific feature, such as (a) addiction(self), (b) affairs(self), (c) aggressive(spouse), (d) child(self), (e) emotional abuse, (f) failed expect, (g) in-laws (self) and (h) in-laws (spouse).

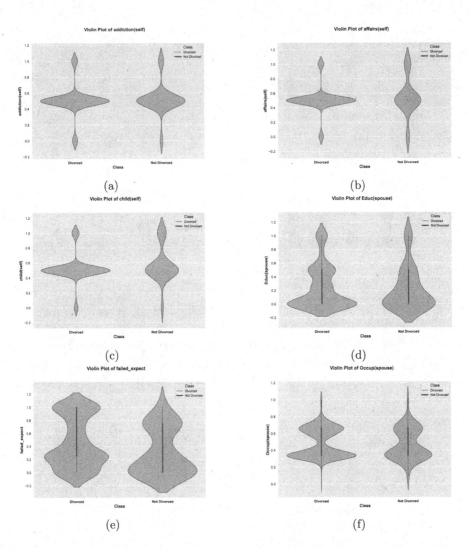

Fig. 4. The violin graph analysis of divorce data. Each subfigure corresponds to a specific feature, (a) addiction(self) among Divorced and Not Divorced Class, (b) affairs(self) among Divorced and Not Divorced Class, (c) child (self) among Divorced and Not Divorced Class, (d) educ (spouse) among Divorced and Not Divorced Class, (e) failed expect among Divorced and Not Divorced Class and (f) occup(spouse) among Divorced and Not Divorced Class.

introducing randomness in feature selection. Recent studies have demonstrated the effectiveness of these techniques in image classification, disease prediction, and cancer detection [4]. These approaches leverage the collective knowledge of multiple decision trees to enhance predictive accuracy across various domains.

4.3 Stacking

Stacking is an ensemble learning technique that combines predictions from multiple base models using a meta-model. It improves prediction performance by leveraging the diverse patterns captured by different models. Subsequent studies have demonstrated its effectiveness in various applications [22].

4.4 Hyperparameter Optimization

In the proposed work, recent optimization algorithms, namely Jaya and Whale Optimization Algorithm (WOA), were employed for hyperparameter optimization. Jaya algorithm, inspired by the concept of improving solutions iteratively, has demonstrated promising results in various optimization tasks [15]. WOA, drawing inspiration from whale hunting behavior, has been widely utilized in optimization problems due to its effectiveness [11]. By leveraging Jaya and WOA, the proposed work fine-tuned model hyperparameters, resulting in enhanced performance and optimized configurations.

Jaya Algorithm. *Initialization:*

1. Initialize a population of candidate solutions: $Y = \{Y_1, Y_2, \ldots, Y_n\}$.
2. Define the objective function: $f(Y_i)$ evaluates solution quality.

Jaya Algorithm Steps:

1. Iterate until termination:
   (a) Calculate fitness $F(Y_i)$ for each solution.
   (b) Identify the best (Y_{best}) and worst (Y_{worst}) solutions.
   (c) Update solutions (Eq. (1) and Eq. (2)):
       − For better solutions:

$$Y_{\text{new}_i} = Y_i + \text{rand}(0,1) \cdot (Y_{\text{best}} - Y_i) \tag{1}$$

   − For the worst solution:

$$Y_{\text{new}_i} = Y_i - \text{rand}(0,1) \cdot (Y_{\text{best}} - Y_i) \tag{2}$$

   (d) Apply constraints if needed.

   Return the best solution found (Y_{best}).

Whale Optimization Algorithm (WOA). *Initialization:*

1. Initialize a population of candidate solutions: $Y = \{Y_1, Y_2, \ldots, Y_n\}$.
2. Define the objective function: $f(Y_i)$ evaluates the fitness of solution Y_i.

WOA Algorithm Steps:

1. Repeat until a termination criterion is met:
   (a) Update the exploration and exploitation coefficient a as in Eq. (3):

$$a = 2 - \frac{2t}{T_{\max}} \tag{3}$$

   where t is the current iteration, and $T_{\max}$ is the maximum number of iterations.
   (b) For each whale Y_i in the population:
   – Randomly select two other whales Y_{rand1} and Y_{rand2}.
   – Update Y_i based on exploration and exploitation phases using Eq. (4)–(7):
   • Exploration phase:

$$D = |C \cdot Y_{\mathrm{rand1}} - Y_i| \tag{4}$$

$$Y_{\mathrm{new1}} = Y_{\mathrm{rand1}} - aD \tag{5}$$

   • Exploitation phase (if $D > 1$):

$$labeleq : 6 Y_{\mathrm{new2}} = A \cdot D - Y_i \tag{6}$$

   • Combine both phases:

$$Y_{\mathrm{new}} = \frac{Y_{\mathrm{new1}} + Y_{\mathrm{new2}}}{2} \tag{7}$$

   – Apply constraints if necessary to ensure that Y_i remains within the feasible solution space.
   (c) Calculate the fitness for each solution in the population: $F(Y_i) = f(Y_i)$ for all i.
   (d) Identify the best solution Y_{best} as in Eq. (8):

$$Y_{\mathrm{best}} = \arg\min(F(Y_i)) \tag{8}$$

   (e) Update the positions of all whales using Eq. (9)):
   – For each whale Y_i:

$$Y_{\mathrm{new}} = Y_{\mathrm{best}} - A \cdot \mathrm{rand}(0,1) \cdot |Y_{\mathrm{best}} - Y_i| \tag{9}$$

   – Apply constraints if necessary to ensure that Y_i remains within the feasible solution space.
Return the best solution found: Y_{best}.

5   Results and Discussion

In our study, we employed a set of well-established evaluation metrics to assess the performance of our classification model. Specifically, we utilized Precision, Recall, and F1-Score, which are widely recognized in classification tasks. Precision provided insights into the accuracy of our model's positive predictions, while Recall gauged its capacity to correctly identify all relevant instances. The F1-Score, a balanced combination of Precision and Recall, offered a balanced measure of our model's classification performance. Additionally, we considered Accuracy, a straightforward yet valuable metric, which quantifies the ratio of correct predictions to the total number of predictions. It's important to note that while Accuracy is informative, it may not be ideal for datasets with imbalanced class distributions. To address this concern, we also employed the Area Under the Receiver Operating Characteristic Curve (AUC-ROC) metric. AUC-ROC allowed us to evaluate the model's ability to differentiate between positive and negative classes, particularly beneficial when dealing with imbalanced datasets and binary classification scenarios.

5.1   Experiments on Adaboost Algorithm

Here we evaluated the performance of the Ada-Boost algorithm on different train-test splits (80:20, 70:30, and 75:25) before and after applying SMOTE for class balancing. We utilized the Jaya and Whale optimization algorithms to tune the hyperparameters of the Ada-Boost algorithm. Various evaluation metrics were employed to measure the performance of the model. As shown in Table 3, Adaboost obtains an accuracy of 97% and 96% while using Jaya and Whale optimization algorithms respectively. Table 4, presents the hyperparameters optimization (HPO) of the AdaBoost algorithm. The conducted experiments on the AdaBoost algorithm reveal that class balancing has a positive impact on the algorithm's performance and demonstrate how fine-tuning hyperparameters enhances the effectiveness of the algorithm.

Table 3. Experiments on Adaboost Algorithm

| | Results of Adaboost Algorithm(%) | | | | | | | |
| | Before Balancing | | | After Balancing | | | After HPO | |
	80/20	75/25	70/30	80/20	75/25	70/30	Jaya	WOA
Accuracy	81	87	83	82	90	83	97	96
Precision	85	89	86	83	90	83	97	96
Recall	81	87	83	82	90	83	97	96
F1 Score	73	81	75	82	90	83	97	96

Table 4. The comparison analysis of AdaBoost Algorithm before and after(Jaya) hyperparameter optimization

Parameters	before HPO	after HPO
n_estimators	10	99
learning rate(sec)	0.001	0.589
random state	3	46
base estimator		Decision Tree
max_depth		2
min samples split		3
resulted accuracy	87	97

5.2 Experiments on Gradient Boosting Algorithm

Different train-test ratios were used to conduct experiments on the gradient boosting algorithm before and after class balancing. As indicated in Table 5, after class balancing, the gradient boosting algorithm achieved an optimal accuracy, precision, recall, and F1 score of 89% with a train-test split ratio of 70:30. Moreover, utilizing Jaya and WOA for hyperparameter optimization resulted in accuracy, precision, recall, and F1 score of 95%

Table 5. Experiments on Gradient Boosting Algorithm

	Results of Gradient Boosting Algorithm (%)							
	Before Balancing			After Balancing			After HPO	
	80/20	75/25	70/30	80/20	75/25	70/30	Jaya	WOA
Accuracy	81	87	83	82	86	89	95	95
Precision	85	89	86	83	86	89	95	95
Recall	81	87	83	82	86	89	95	95
F1 Score	73	81	75	82	86	89	95	95

5.3 Experiments on Bagging Algorithm

The experiments conducted on the bagging algorithm demonstrated that its highest performance before class balancing was accuracy of 87%, precision of 89%, recall of 87%, and an F1 score of 81% when using a train-test ratio of 75:25. After class balancing and maintaining the same train-test ratio, the performance improved, achieving 92% for all four measures. Furthermore, by employing Jaya hyperparameter optimization with various parameters for the bagging algorithm, an impressive 95% accuracy, precision, recall, and F1 score were achieved. Table 6, presents the experimental results of the Bagging algorithm.

Table 6. Experiments on Bagging Algorithm

	Results of Bagging Algorithm (%)							
	Before Balancing			After Balancing			After HPO	
	80/20	75/25	70/30	80/20	75/25	70/30	Jaya	WOA
Accuracy	81	87	83	82	92	83	95	89
Precision	85	89	86	83	92	83	95	89
Recall	81	87	83	82	92	83	95	89
F1 Score	73	81	75	82	92	83	95	89

5.4 Experiments on Stacking Algorithm

As presented in Table 7, the stacking algorithm attains its optimal performance using a train-test split ratio of 75:25, with accuracy, precision, recall, and F1 scores reaching 87%, 89%, 87%, and 81%, respectively, before class balancing. After class balancing, the performance significantly improved, reaching 93% for all four measures. Additionally, by applying Jaya and WOA optimizations for hyperparameter tuning of the algorithm, 94% was achieved for the performance measures.

Table 7. Experiments on Stacking Algorithm

	Results of Stacking Algorithm (%)							
	Before Balancing			After Balancing			After HPO	
	80/20	75/25	70/30	80/20	75/25	70/30	Jaya	WOA
Accuracy	81	87	83	93	91	92	94	94
Precision	85	89	86	93	91	92	94	94
Recall	81	87	83	93	91	92	94	94
F1 Score	73	81	75	93	91	92	94	94

5.5 Experiments on XGBoost Algorithm

Table 8, demonstrates that the performance of the XGBoost algorithm significantly improves after class balancing, achieving an impressive 94% for all four performance evaluation metrics. Moreover, when applying both Jaya and WOA optimizations for hyperparameter tuning and considering various parameters of XGBoost, the algorithm's performance further elevates to 95%, taking into account diverse parameters that fine-tune the XGBoost algorithm.

Table 8. Experiments on XGBoost Algorithm

| | Results of XGBoost Algorithm(%) | | | | | | | |
| | Before Balancing | | | After Balancing | | | After HPO | |
	80/20	75/25	70/30	80/20	75/25	70/30	Jaya	WOA
Accuracy	87	84	84	91	94	92	95	95
Precision	85	89	81	91	94	92	95	95
Recall	82	87	84	91	94	92	95	95
F1 Score	75	81	78	91	94	92	95	95

5.6   Experiments on Random Forest Algorithm

Based on the conducted experiments on the random forest algorithm shown in Table 9, it is evident that addressing the class imbalance leads to a notable improvement in the model's performance. The model achieves impressive accuracy, precision, recall, and F1 score values of 92% for both the 70:30 and 80:20 data split ratios. Furthermore, by utilizing both Jaya and WOA optimization algorithms for hyperparameter tuning of the random forest algorithm, the model's performance reaches an optimized level of 95% for all four evaluation metrics.

Table 9. Experiments on Random Forest Algorithm

| | Results of Random Forest Algorithm | | | | | | | |
| | Before Balancing | | | After Balancing | | | After HPO | |
	80/20	75/25	70/30	80/20	75/25	70/30	Jaya	WOA
Accuracy	80	85	83	92	91	92	95	95
Precision	66	75	75	92	91	92	95	95
Recall	80	85	83	92	91	92	95	95
F1 Score	72	80	76	92	91	92	95	95

5.7   Roc Curve of Ensemble Algorithms After Jaya Optimization

The ROC curve analysis is a method used to assess the performance of classification algorithms. The ROC curve analysis evaluated several classification algorithms, with AdaBoost, Random Forest (RF), Gradient Boosting (GB), and XGBoost achieving a remarkable AUC value of 0.99 as indicated in Fig. 5. This indicates their exceptional ability to accurately classify divorce. However, the AUC curve indicates the XGBoost has outperformed the rest of the algorithms.

In contrast, stacking and bagging algorithms attained slightly lower AUC values of 0.93 and 0.98, respectively. While not as high as the top-performing algorithms, these values still reflect their effective classification performance.

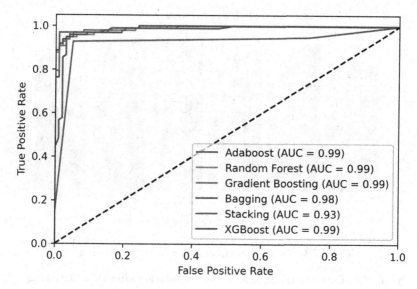

Fig. 5. ROC curve of ensemble algorithms after class balancing and Jaya Optimization

5.8   Accuracy Comparison After Feature Selection Using Genetic Algorithm

In our study, we employed a genetic algorithm to carefully select 28 features from the initial set of 64 features. These selected features were utilized to build predictive models, and we evaluated the models across three distinct train-validation-test split ratios.

Upon analyzing the results, it became evident that the Bagging algorithm outperformed the other methods. Bagging algorithm achieved an impressive accuracy rate of 94.88% using 70:15:15, 94.90% using both 75:15:10 and 80:10:10 train-validation-test split ratios as witnessed from Fig. 6.

5.9   K-Fold Cross-Validation Results of Ensemble Algorithms After Class Balancing

In our cross-validation experiments with k values of 5, 7, and 10, bagging emerged as the top-performing algorithm. It consistently achieved a high level of accuracy, precision, recall, and F1 score, all at 94.98%. This impressive performance demonstrates bagging's robustness and its ability to make accurate predictions, setting it apart as a strong choice among the algorithms considered. This is based on the results shown in Table 10.

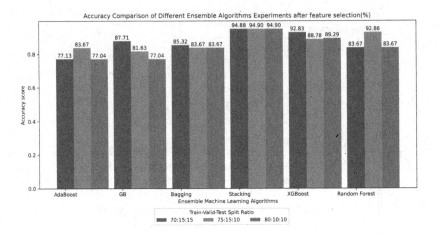

Fig. 6. Accuracy comparison after Feature selection

Table 10. Experiments on k-fold Cross-Validation after class Balancing

Average Results	k-fold Cross-Validation after class Balancing(%)				
	AdaBoost	RF	Stacking	Bagging	XGBoost
	K = 10	K = 5	K = 10	K = 7	K = 10
Accuracy	82.48	94.57	93.95	94.98	92.21
Precision	82.90	94.57	94.01	94.98	92.22
Recall	82.48	94.57	93.95	94.98	92.21
F1 Score	82.42	94.57	93.95	94.98	92.21

5.10 Training Vs Validation Vs Testing Accuracy and Logarithmic Loss After Jaya Optimization

According to Fig. 7 as in the accuracy plot, we examine how well various ensemble machine learning models classify data. Training accuracy is consistently high across all models, indicating effective learning on the training data. While high training accuracy is expected, the primary goal is good performance on validation data, which simulates real-world situations. Minor variations in validation accuracy suggest differences in generalization ability. The results indicate that overfitting is not a significant concern, as there's little difference between training and validation accuracy. 'Gradient Boosting' and 'XGBoost' stand out with the highest validation accuracy.

The loss plot explores into the trade-off between model complexity and predictive power. Training loss, measuring how well models fit the training data, is low across all models. Validation loss, reflecting generalization to new data, is also low with minor variations. Consistent training and validation loss values indicate no significant overfitting. Lower loss values signify better predictive power. 'Random Forest' demonstrates a balance between low training and validation loss.

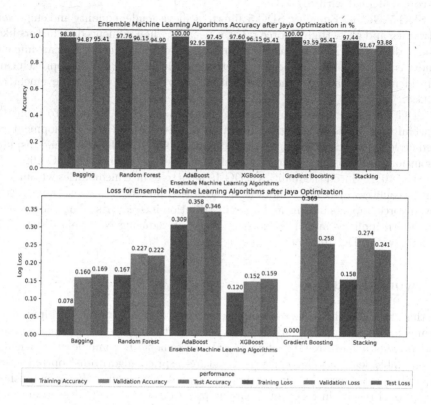

Fig. 7. Accuracy and Log loss comparisons after Jaya Optimization

6   Contributions of Studying Divorce to Sustainable Development Goals (SDGs)

Studying divorce contributes to several Sustainable Development Goals (SDGs) by addressing social, economic, and well-being aspects of individuals and societies.

SDG 1: No Poverty: SDG 1 seeks to end poverty in all its forms. Divorce can have socio-economic consequences, particularly for women who may face financial challenges after divorce. Understanding and predicting divorce patterns can inform poverty reduction strategies by addressing the economic vulnerabilities associated with divorce.

SDG 3: Good Health and Well-being: SDG 3 focuses on promoting well-being for all ages. Divorce can have significant emotional and mental health impacts on individuals and children. Accurate divorce prediction models can lead to early intervention and support, contributing to improved mental and emotional health outcomes aligned with SDG 3.

SDG 5: Gender Equality: SDG 5 aims to achieve gender equality and empower all women and girls. High divorce rates, especially when influenced by factors like gender disparities in asset distribution and custody arrangements, can impact gender equality. Using advanced predictive models like this hybrid approach can help identify and address such disparities, contributing to the achievement of SDG 5.

SDG 16: Peace, Justice, and Strong Institutions: SDG 16 seeks to promote peaceful and inclusive societies. Research on divorce and the development of predictive models can contribute to the creation of fair legal systems and dispute resolution mechanisms, fostering peaceful societies as outlined in SDG 16.

SDG 10: Reduced Inequalities: SDG 10 aims to reduce inequalities within and among countries. Divorce studies using advanced predictive models can identify how divorce impacts income and asset distribution inequalities. This information can inform policies and interventions aimed at reducing economic disparities, aligning with SDG 10.

7 Conclusion

In this study, the hybridization of ensemble learning and nature-inspired algorithms for divorce prediction holds the potential to improve prediction accuracy and provide a more reliable divorce risk assessment. It combines the benefits of ensemble learning's diversity and nature-inspired algorithms' optimization capabilities to create a robust and efficient predictive model. We accomplished exceptional results in terms of accuracy, precision, recall, and F1 score for several algorithms by employing the Jaya optimization algorithm. Specifically, we achieved 95% for RF, XGBoost, Gradient Boosting, Bagging, and 94% and 97% for Stacking and AdaBoost, respectively. Furthermore, when applying the WOA algorithm, we obtained 95% accuracy, precision, recall, and F1 score for Gradient Boosting, Random Forest, and XGBoost algorithms. For Stacking, we achieved 94%, while for AdaBoost, we attained 96% accuracy. Additionally, the Bagging algorithm resulted in 89% accuracy, 90% precision, 89% recall, and 89% F1 score. Based on these findings, we can conclude that the AdaBoost algorithm exhibits the best performance when both Jaya and WOA are utilized for hyperparameter optimization. To enhance machine learning models, we recommend to consider: 1) Exploring diverse nature-inspired algorithms for optimization. 2) Leveraging advanced feature engineering techniques for richer insights. 3) Prioritizing larger and more diverse datasets to improve model training and reliability. Moreover, by incorporating explainable techniques and methods, you can enhance the transparency of model's decision-making process. These strategies empower the development of highly effective models to tackle real-world challenges.

References

1. AL-Behadili, H.N.K., Ku-Mahamud, K.R.: Hybrid k-nearest neighbour and particle swarm optimization technique for divorce classification. Int. J. Adv. Sci. Eng. Inf. Technol. **11**(4), 1447–1454 (2021). https://doi.org/10.18517/ijaseit.11.4.14868, www.ijaseit.insightsociety.org/index.php?option=com_content&view=article&id=9&Itemid=1&article_id=14868
2. Amato, P.R., Previti, D.: People's reasons for divorcing: gender, social class, the life course, and adjustment. J. Fam. Issues **24**(5), 602–626 (2003). https://doi.org/10.1177/0192513X03254507
3. Asfaw, L.S., Alene, G.D.: Marital dissolution and associated factors in hosanna, southwest Ethiopia: a community-based cross-sectional study. BMC psychology **11**(1), 1–10 (2023). https://doi.org/10.1186/s40359-023-01051-3
4. Breiman, L.: Random forests. Mach. Learn. **45**, 5–32 (2001). https://doi.org/10.1023/A:1010933404324
5. Chen, T., Guestrin, C.: Xgboost: A scalable tree boosting system. In: Proceedings of the 22nd ACM Sigkdd International Conference on Knowledge Discovery and Data Mining, pp. 785–794 (2016). https://doi.org/10.1145/2939672.2939785
6. Cherlin, A.J.: Demographic trends in the united states: a review of research in the 2000s. J. Marriage Fam. **72**(3), 403–419 (2010). https://doi.org/10.1111/j.1741-3737.2010.00710.x
7. Cleveland, W.S.: Visualizing Data, vol. 36. Hobart Press, Cleveland (1993). https://doi.org/10.2307/1269376
8. Dagnew, G.W., Asresie, M.B., Fekadu, G.A., Gelaw, Y.M.: Factors associated with divorce from first union among women in Ethiopia: further analysis of the 2016 Ethiopia demographic and health survey data. PLoS ONE **15**(12), e0244014 (2020). https://doi.org/10.1371/journal.pone.0244014
9. Freund, Y., Schapire, R.E., et al.: Experiments with a new boosting algorithm. In: ICML, vol. 96, pp. 148–156. Citeseer (1996)
10. Friedman, J.H.: Greedy function approximation: a gradient boosting machine. Ann. Stat. **29**, 1189–1232 (2001). https://doi.org/10.1214/AOS/1013203451
11. Gharehchopogh, F.S., Gholizadeh, H.: A comprehensive survey: whale optimization algorithm and its applications. Swarm Evol. Comput. **48**, 1–24 (2019). https://doi.org/10.1016/J.SWEVO.2019.03.004
12. Karney, B.R., Bradbury, T.N.: The longitudinal course of marital quality and stability: a review of theory, methods, and research. Psychol. Bull. **118**, 3–34 (1995). https://doi.org/10.1037/0033-2909.118.1.3
13. Kuncheva, L.I.: Combining Pattern Classifiers: Methods and Algorithms. Wiley, United Kingdom (2014). https://doi.org/10.1002/0471660264
14. Ranjitha, P., Prabhu, A.: Improved divorce prediction using machine learning-particle swarm optimization (PSO). In: 2020 International Conference for Emerging Technology (INCET), pp. 1–5. IEEE (2020). https://doi.org/10.1109/INCET49848.2020.9154081
15. Rao, R.: Jaya: a simple and new optimization algorithm for solving constrained and unconstrained optimization problems. Int. J. Ind. Eng. Comput. **7**(1), 19–34 (2016). https://doi.org/10.5267/j.ijiec.2015.8.004
16. Rokach, L.: Ensemble-based classifiers. Artif. Intell. Rev. **33**, 1–39 (2010). https://doi.org/10.1007/s10462-009-9124-7
17. Sadiq Fareed, M.M., et al.: Predicting divorce prospect using ensemble learning: support vector machine, linear model, and neural network. Comput. Intell. Neurosci. **2022** (2022). https://doi.org/10.1155/2022/3687598

18. Sahoo, K., Samal, A.K., Pramanik, J., Pani, S.K.: Exploratory data analysis using python. Int. J. Innov. Technol. Explor. Eng. **8**(12), 4727–4735 (2019). https://doi.org/10.35940/ijitee.l3591.1081219

19. Shankhdhar, A., Gupta, T., Gautam, Y.V.: Divorce prediction scale using improvised machine learning techniques. In: Suma, V., Chen, J.I.-Z., Baig, Z., Wang, H. (eds.) Inventive Systems and Control. LNNS, vol. 204, pp. 777–788. Springer, Singapore (2021). https://doi.org/10.1007/978-981-16-1395-1_57

20. Simanjuntak, M., Muljono, M., Shidik, G.F., Zainul Fanani, A.: Evaluation of feature selection for improvement backpropagation neural network in divorce predictions. In: 2020 International Seminar on Application for Technology of Information and Communication (iSemantic), pp. 578–584 (2020). https://doi.org/10.1109/iSemantic50169.2020.9234297

21. Tilson, D., Larsen, U.: Divorce in Ethiopia: the impact of early marriage and childlessness. J. Biosoc. Sci. **32**(3), 355–372 (2000). https://doi.org/10.1017/S0021932000003552

22. Wolpert, D.H.: Stacked generalization. Neural Netw. **5**(2), 241–259 (1992). https://doi.org/10.1016/S0893-6080(05)80023-1

23. Yibre, A.M., Koçer, B.: Semen quality predictive model using feed forwarded neural network trained by learning-based artificial algae algorithm. Eng. Sci. Technol. Int. J. **24**(2), 310–318 (2021). https://doi.org/10.1016/j.jestch.2020.09.001

24. Yöntem, M.K., Kemal, A., Ilhan, T., Kiliçarslan, S.: Divorce prediction using correlation based feature selection and artificial neural networks. Nevşehir Hacı Bektaş Veli Üniversitesi SBE Dergisi **9**(1), 259–273 (2019)

Author Index

A

Abate Tefera, Yalemsew   I-262
Abate, Solomon Teferra   I-237
Abdalaziz, Ahmed   I-73
Abebe, Abel   I-250
Abraham, Akalu   I-60
Adejumo, Abduljaleel   I-88
Aga, Rosa Tsegaye   I-237, I-250
Andargie, Fitsum Assamnew   I-101
Asmare, Habtamu Shiferaw   I-203
Azamuke, Denish   II-62

B

Bainomugisha, Engineer   II-62
Biru, Bereket Hailu   II-129
Blaauw, Dewald   II-83, II-107, II-151

C

Chikotie, Taurai T.   II-197
Choudja, Pauline Ornela Megne   I-88
Churu, Matida   II-151

D

Debelee, Taye Girma   I-237, I-250
Demissie, Dereje Degeffa   I-101
Demma Wube, Hana   II-30
Deneke, Bereket Siraw   I-250

E

Eissa, Khaled   II-219

F

Fayiso Weldesellasie, Firesew   II-30

G

Gachena, Worku   I-237, I-250
Gebretatios, Daniel Tesfai   I-221

Ghebraeb, Rutta Fissehatsion   I-221
Ghebregiorgis, Bereket Desbele   I-221
Girma Debelee, Taye   I-3, I-46, II-30
Gizachew, Beakal   I-176

H

Habtamu, Robbel   I-176
Hailemariam Woldegebreal, Dereje   I-262
Heyl, Isabelle   II-83

J

Jifara, Worku   I-154

K

Katarahweire, Marriette   II-62
Keleta, Mussie Kaleab   I-221
Kidane, Mebrahtu Fisshaye   I-221
Kizza, Samuel   II-62

L

Lambamo, Wondimu   I-154

M

Megersa Ayano, Yehualashet   I-3, I-46
Mekonnen, Rahel   I-237, I-250
Melese Motuma, Mersibon   I-3
Melese, Solomon Zemene   II-129
Menebo, Muluken   II-171
Menzel, Wolfgang   I-141
Merkebu, Solomon   I-237
Meshesha, Million   I-60
Mohamed, Yesuf   I-141
Mohammed Hussien, Habib   I-262
Mulat, Abel   I-237, I-250
Mulat, Ashenafi   I-237, I-250
Muleesi Businge, Joshua   II-62
Mulugeta, Hiwot   I-237, I-250

T. G. Debelee et al. (Eds.): PanAfriConAI 2023, CCIS 2069, pp. 265–266, 2024.
https://doi.org/10.1007/978-3-031-57639-3

N
Negash, Lebsework   II-171
Nyunga Mpinda, Berthine   II-3

O
Olawale Awe, Olushina   II-3
Oluchukwu Njoku, Anthonia   II-3
Opio, Chrisostom   II-62

S
Sahle, Kalkidan A.   II-242
Samuel, Mesay   I-250
Schwenker, Friedhelm   I-73, II-219
Shiferaw, Dereje   II-171
Sisay Hailu, Samuel   I-46
Sow, Binta   I-88
Srinivasagan, Ramasamy   I-154

T
Tachbelie, Martha Yifiru   I-237
Taye Zewde, Elbetel   I-3

T
Tekle, Yonatan Yosef   I-221
Terefe Debella, Tsegamlak   I-262
Tsehay Demis, Betimihirt Getnet   I-117
Tuse, Misganu   I-60

V
Vincent Banda, Takudzwa   II-107

W
Watson, Bruce W.   II-107, II-197
Watson, Bruce   II-83, II-151
Watson, Liam R.   II-197
Wolde Feyisa, Degaga   I-3, I-46

Y
Yefou, Uriel Nguefack   I-88
Yibre, Abdulkerim M.   I-117, II-242
Yibre, Abdulkerim Mohammed   I-203

Z
Zekarias Esubalew, Sintayehu   II-30

Printed in the United States
by Baker & Taylor Publisher Services